TAMBORA

TAMBORA

THE ERUPTION THAT CHANGED THE WORLD

GILLEN D'ARCY WOOD

PRINCETON UNIVERSITY PRESS

PRINCETON AND OXFORD

Copyright ©2014 by Princeton University Press
Published by Princeton University Press, 41 William Street, Princeton, New Jersey 08540
In the United Kingdom: Princeton University Press, 6 Oxford Street, Woodstock, Oxfordshire OX20 1TW

press.princeton.edu

Jacket Art: *Weymouth Bay*, 1816 (oil on canvas), John Constable (1776–1837). Location: The Victoria & Albert Museum, London, UK. Courtesy of The Bridgeman Art Library.

Fourth printing, and first paperback printing, 2015

Cloth ISBN: 978-0-691-15054-3

Paper ISBN: 978-0-691-16862-3

The Library of Congress has cataloged the cloth edition as follows:
Wood, Gillen D'Arcy.
Tambora : the eruption that changed the world / Gillen D'Arcy Wood.
pages cm
Includes bibliographical references and index.
ISBN 978-0-691-15054-3 (hardcover : acid-free paper) 1. Tambora, Mount (Indonesia)—Eruption, 1815. 2. Weather—Effect of volcanic eruptions on—History—19th century. 3. Volcanoes—Environmental aspects—History—19th century. 4. Climatology—Observations—History—19th century. I. Title.
QC981.8.V65W66 2014
363.34'95—dc23
2013021152

British Library Cataloging-in-Publication Data is available

This book has been composed in Sabon Next LT Pro

Printed on acid-free paper.

Printed in the United States of America

10 9 8 7 6 5 4

To the memory of Bess, Linnell, Monica, and Bessie—
And to Nancy, and a climate-stable future for our children

A fearful hope was all the world contained . . .
The brows of men by the despairing light
Wore an unearthly aspect, as by fits
The flashes fell upon them; some lay down
And hid their eyes and wept . . .
And others hurried to and fro, and fed
Their funeral piles with fuel, and looked up
With mad disquietude on the dull sky,
The pall of a past world

LORD BYRON, "DARKNESS" (1816)

CONTENTS

List of Illustrations   xi

Note on Measurements   xv

INTRODUCTION
Frankenstein's Weather   1

ONE
The Pompeii of the East   12

TWO
The Little (Volcanic) Ice Age   33

THREE
"This End of the World Weather"   45

FOUR
Blue Death in Bengal   72

FIVE
The Seven Sorrows of Yunnan   97

SIX
The Polar Garden   121

SEVEN
Ice Tsunami in the Alps   150

EIGHT
The Other Irish Famine   171

NINE
Hard Times at Monticello   199

EPILOGUE
Et in Extremis Ego   229

Acknowledgments   235

Notes   237

Bibliography   259

Index   281

ILLUSTRATIONS

FIGURES

FIGURE 0.1
Map of global sulfate deposition   3

FIGURE 0.2
Caspar David Friedrich, *Ships in the Harbor* (1816)   4

FIGURE 0.3
Tambora caldera   6

FIGURE 0.4
Tambora caldera (aerial view)   7

FIGURE 1.1
Map of the East Indies (nineteenth century)   13

FIGURE 1.2
Map of Sumbawa   14

FIGURE 1.3
Diagram of plinian explosion and pyroclastic flow   19

FIGURE 1.4
Tambora eruption timeline   22

FIGURE 1.5
Map of Tambora ash fallout   23

FIGURE 1.6
Artifacts recovered from Tambora village   26

FIGURE 1.7
Portrait of Sir Stamford Raffles (1817)   28

FIGURE 1.8
Javan landscape (nineteenth century)   31

FIGURE 2.1
J.M.W. Turner, *Mt. Vesuvius in Eruption* (1812)   35

FIGURE 2.2
Sulfate deposition in ice core samples showing 1809 and 1815
eruptions   38

FIGURE 3.1
Portrait copy of Mary Shelley (1816)   47

FIGURE 3.2
Portrait of Percy Bysshe Shelley (1819)   47

FIGURE 3.3
Diagram showing creation of volcanic sulfate aerosols   48

FIGURE 3.4
John Constable, *Weymouth Bay* (1816)   50

FIGURE 3.5
Annual frequency of gale days at Edinburgh (1789–1988)   53

FIGURE 3.6
Lord Byron at the Villa Diodati   68

FIGURE 4.1
Historical map of north-central India, scene of Lord Hastings's
Maratha campaign and the cholera outbreak of November 1817   73

FIGURE 4.2
A View of Erich above the River Betwah (1817)   76

FIGURE 4.3
Synoptic map of South Asian monsoon   80

FIGURE 4.4
William Hodges, *The Ghauts at Benares* (1787)   85

FIGURE 4.5
James Baillie Fraser, *Calcutta Bazaar* (1824–26)   91

FIGURE 4.6
Map of global cholera spread (nineteenth century)   92

FIGURE 5.1
Map of China agriculture (Qing dynasty)   100

FIGURE 5.2
Chinese rice fields (mid-1840s)   103

FIGURE 5.3
Chinese mother bringing her children to market for sale
(mid-1870s)   112

FIGURE 5.4
Chinese opium den (mid-1840s)   118

FIGURE 6.1
Portrait of Sir John Barrow (circa 1810)   124

FIGURE 6.2
Portrait of William Scoresby Jr. (1821)   126

FIGURE 6.3
Scoresby's "Marine Diver"   136

FIGURE 6.4
Map showing global thermohaline circulation   137

FIGURE 6.5
Global drought post–Pinatubo eruption   139

FIGURE 6.6
Portrait of Captain William Edward Parry (1820)   143

FIGURE 6.7
Discovery of Franklin expedition remains (1861)   145

FIGURE 7.1
J.M.W. Turner, *Mer de Glace* (1812) 156

FIGURE 7.2
Portrait of Ignace Venetz (1826) 158

FIGURE 7.3
Map of Val de Bagnes and Giétro glacier dam, 1818 160

FIGURE 8.1
The Black Prophet (title page) 175

FIGURE 8.2
Synoptic map of low-pressure system over British Isles, July 1816 177

FIGURE 8.3
Diagram showing typhus transmission 184

FIGURE 8.4
Illustration, *The Black Prophet* 196

FIGURE 9.1
Map of New England snowline, June 6–7, 1816 202

FIGURE 9.2
Historical graph, New England growing seasons 204

FIGURE 9.3
Portrait of Thomas Jefferson (1821) 207

FIGURE 9.4
Portrait of Comte de Buffon (1753) 208

FIGURE 9.5
Transatlantic grain prices, post-Tambora period 222

FIGURE 9.6
View of Monticello (1825) 223

FIGURE E.1
Nicolas Poussin, *Et in Arcadia Ego* (1639) 232

NOTE ON MEASUREMENTS

This book deals much with science and the history of science, but I have not maintained a "scientific" adherence to metrical units of measurement. Instead I have been guided principally by context. Where the historical or cultural setting seems appropriate, I have used imperial measures of distance as well as the Fahrenheit temperature scale.

TAMBORA

INTRODUCTION

FRANKENSTEIN'S WEATHER

The War of Independence between Britain and America provisionally ended with the Treaty of Paris in December 1783. But official ratification of the peace accord was delayed for months by a mix of political logistics and persistent bad weather. The makeshift U.S. capital in Annapolis, Maryland, was snowbound, preventing assembly of congressional delegates to ratify the treaty, while storms and ice across the Atlantic slowed communications between the two governments. At last, on May 13, 1784, Benjamin Franklin, wrangling matters in Paris, was able to send the treaty, signed by King George himself, to the Congress.

Even while scrambling to bring the warring parties to terms, Franklin—tireless and mercurial—found time to reflect on the altered climate of 1783–84 that had played such a complicating role in recent events. "There seems to be a region high in the air over all countries where it is always winter," he wrote. But perhaps the "universal fog" and cold that had descended from the atmosphere to blanket all Europe might be attributed to volcanic activity, specifically an eruption in nearby Iceland.[1]

Franklin's "Meteorological Imaginations and Conjectures" amounts to no more than a few pages of disconnected thoughts, scribbled amid a high-stakes diplomatic drama. The paper's unlikely fame as a scientific document rests on its being the first published speculation on the link between volcanism and extreme weather. Franklin hastily sent his paper on meteorology to Manchester, where the local Philosophical Society had awarded him honorary membership. On December 22, 1784, the president of the society rose to speak on Franklin's behalf. No doubt dismayed at the paper's thinness, he had no choice but to read the "conjectures" of the society's celebrated new member to the crowded assembly. There, in a freezing Manchester public hall, the theory that volcanic eruptions are capable of wreaking climate havoc was given its first public utterance.

No one believed it for a moment. Even as the hall emptied, Franklin's idea had entered the long oblivion of prematurely announced truths. But, of course, he was right. The eruption of the Iceland volcano Laki in June 1783 brought abrupt cooling, crop failures, and misery to Europe the following year, and created dangerously icy conditions for Atlantic shipping. Even so, Laki did not go global. Latitude is critical to the relation between volcanic eruptions and climate. As a high northern volcano, Laki's ejecta did not penetrate the trans-hemispheric currents of the planet's climate system, and its meteorological impacts were confined to the North Atlantic and Europe.

Two hundred years ago, no one—not even Benjamin Franklin—had grasped the potential global impact of volcanic emissions from the *tropics*, where, two decades after Laki, planet Earth's greatest eruption of the millennium took place. When Mount Tambora—located on Sumbawa Island in the East Indies—blew itself up with apocalyptic force in April 1815, no one linked that single, barely reported geological event with the cascading worldwide weather disasters in its three-year wake.

Within weeks, Tambora's stratospheric ash cloud had circled the planet at the equator, from where it embarked on a slow-moving sabotage of the global climate system at all latitudes. Five months after the eruption, in September 1815, meteorological enthusiast Thomas Forster observed strange, spectacular sunsets over Tunbridge Wells near London. "Fair dry day," he wrote in his weather diary—but "at sunset a fine red blush marked by diverging red and blue bars."[2] Artists across Europe took note of the changed atmosphere. William Turner drew vivid red skyscapes that, in their coloristic abstraction, seem like an advertise-

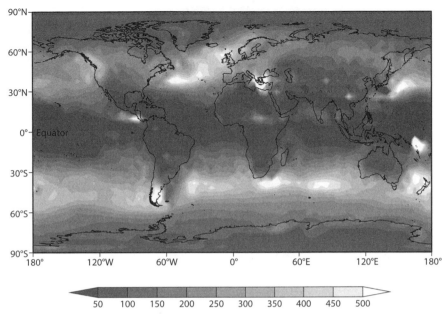

50   100   150   200   250   300   350   400   450   500

FIGURE 0.1. This 2007 model of Tambora's sulfate cloud shows its global reach, with a band of high aerosol concentration at mid- to high latitudes in both hemispheres, notably over the North Atlantic and Western Europe. This model situates the volcanic cloud in the stratosphere, 24–32 kilometers above the Earth. (Chaochao Gao, "Atmospheric Volcanic Loading Derived from Bipolar Ice Cores: Accounting for the Spatial Distribution of Volcanic Deposition," *Journal of Geophysical Research* 112 [2007]: D09109; © American Geophysical Union).

ment for the future of art. Meanwhile, from his studio on Greifswald Harbor in Germany, Caspar David Friedrich painted a sky with a chromic density that—one scientific study has found—corresponds to the "optical aerosol depth" of the colossal volcanic eruption that year.[3]

Forster, Turner, and Friedrich—all committed skywatchers—saw the imprint of major atmospheric changes in the North Atlantic. But neither Forster's London sky "on fire" in September 1815 nor the nearly three years of destructive global cooling that ensued inspired anyone to the realization that a faraway volcanic eruption had caused it all. Not until the Cold War—and the development of meteorological instruments to measure nuclear fallout—did scientists begin study of the atmospheric residency of volcanic aerosols. The sun-blocking dust veil

FIGURE 0.2. Caspar David Friedrich, *Ships in the Harbor* (1816). Sanssouci Palace, Potsdam. (© Erich Lessing/Art Resource, NY).

of a major eruption, it was concluded, might linger above the earth for up to three years. Two centuries after Franklin's first tentative speculations, the geophysical chain linking volcanism and climate could at last be proven.

I dwell on this point for good reason. The formidable, occasionally mind-bending challenge in writing this book has been to trace cataclysmic world events the cause of which the historical actors themselves were ignorant. Generations of historians since have done little better. The Tamboran climate emergency followed hard upon the devastations of the Napoleonic Wars and has always remained in the shadows of that epochal conflict. Out of sight and out of mind, Tambora was the volcanic stealth bomber of the early nineteenth-century. Be it the retching cholera victim in Calcutta, the starving peasant children of Yunnan or County Tyrone, the hopeful explorer of a Northwest Passage through the Arctic Ocean, or the bankrupt land speculator in Baltimore, the world's residents were oblivious to the volcanic drivings of their fate. Equally challenging for me as an environmental historian has been to capture the physically remote relation between cause and effect in measuring Tambora's impact on the global commons of the nineteenth century. Volcanic strife traveled great distances and via obscure agents. But it is only by tracing such "teleconnections"—a guiding principle of today's climate and ecological sciences—that the *worldwide* tragedy of Tambora can be rescued from its two-century oblivion.

Climate change is hard to see and no less difficult to imagine. After a day's climb through the dense forests of Sumbawa Island, drenched in tropical rains, I almost didn't succeed in seeing at firsthand the great Tambora's evacuated peak. Then, at daybreak on the second morning, the clouds suddenly lifted, and we were able to complete our ascent along the treeless ridges. Nearing the summit, we clambered over flat pool tables of tan, serrated rock and left our boot prints in the black volcanic sand. Almost without warning, we found ourselves at the rim of a great inverted dome of earth, with sheer rock walls leading down to a pearl-green lake a kilometer below. My camera whirred as puff-clouds of sulfur performed lazy inversions in the still, separate universe

FIGURE 0.3. Tambora's caldera. The morning this photograph was taken (March 3, 2011), the mountain rumbled and the odor of sulfur was palpable. A few weeks later, the volcano began belching ash and smoke. By September that year, Indonesian seismologists had ordered evacuation of the surrounding area. Volcanologists do not expect an imminent eruption, however, on account of the geologically recent 1815 event. (Author photo).

of Tambora's yawning caldera. Its six-kilometer diameter might as well have been a thousand. My swimming eyes performed no better than my camera in taking measure of the volcano's unhealed intestinal canyon, let alone in imagining its once pristine peak a mile above me in what was now open sky. Sleepless and damp in our tents the night before, we had felt rumblings from deep in the earth. Now, we smelled the distinct odor of sulfur in the morning air. Looking down for a moment to recover my senses, I realized I was standing on a sponge-like rock that, but a blink before in geological time, had been adrift among the brewing magma of Tambora's subterranean chamber.

Gazing out across that dizzying crater, I felt no better equipped than pioneer meteorologist Thomas Forster in 1815 to grasp the catastrophic

FIGURE 0.4. An aerial view of Tambora's caldera taken from the International Space Station shows its terrific, lunar-like dimensions. (NASA).

impact of a single mountain's explosion on the history of the modern world. It was a calm sunrise. The straits of Teluk Saleh to the south came into view over the treetops, its postcard blue waters dotted with islands in the milky sunlight. Stretching behind us, the forests of the Sanggar peninsula appeared at perfect peace. Did an event of world-changing violence truly happen here? Like the shivering audience in that Manchester hall two centuries ago, trying to make sense of Franklin's ramblings about cold weather and an Iceland volcano, I could hardly believe in Tambora's global reach.

It has taken five years of research into the science of volcanism and climate, collaboration with scholars across many disciplines, and much dogged detective work to remake that morning's ascent of Tambora in my imagination: to articulate, in book-length form, the years-long impact of the massive 1815 eruption on the world in the critical period after the Napoleonic Wars. Unlike Benjamin Franklin and Thomas Forster, I have had the advantage of modern scientific instruments and data through which to "see" the otherwise invisible teleconnections linking tropical eruptions, climate change, and human affairs. Climbing Tambora, by this route, one could not mistake its greatness.

Tambora belongs to a dense volcanic cluster along the Sunda arc of the Indonesian archipelago. This east-west ridge of volcanoes is a segment, in turn, of the much larger Ring of Fire, a hemisphere-girdling string of volcanic mountains bordering the Pacific Ocean from the southern tip of Chile, to Mount St. Helens in Washington State, to picturesque Mount Fuji in Japan, to Tambora's near neighbor Krakatau, due to explode into global fame in 1883. Along its almost 40,000 kilometers length, the Ring of Fire boasts lofty, cone-topped volcanoes located exclusively on coastlines and islands. Tambora sits some 330 kilometers north of a tectonic ridge in the trans-Pacific Ring of Fire known as the Java Trench, which marks a curvilinear course south of the islands of Sumbawa and its neighbors Lombok and Sumba.

After perhaps a thousand years' dormancy, Tambora's devastating evacuation and collapse in April 1815 required only a few days. It was the concentrated energy of this event that was to have the greatest human impact. By shooting its contents into the stratosphere with such biblical force, Tambora ensured its volcanic gases reached sufficient height to seriously disable the seasonal rhythms of the global climate system, throwing human communities worldwide into chaos. The sun-dimming stratospheric aerosols produced by Tambora's eruption in 1815 spawned the most devastating, sustained period of extreme weather seen on our planet in perhaps thousands of years.

A dramatic story unto itself. But a more urgent motivation has driven my history of Tambora. The great Sumbawan volcano is the *most recent* volcanic eruption to have had a dramatic impact on global climate. Considered on a geological timescale, Tambora stands almost insistently near to us, begging to be studied. On the eve of Tambora's bicentenary and facing multiplying extreme weather crises of our own, its eruption looms as the richest case study we have for understanding how abrupt shifts in climate affect human societies on global scales and decadal time frames. The Tambora climate emergency of 1815–18 offers us a rare, clear window onto a world convulsed by weather extremes, with human communities everywhere struggling to adapt to sudden, radical shifts in temperatures and rainfall, and a flow-on tsunami of famine, disease, dislocation, and unrest.

For three years following Tambora's explosion, to be alive, almost anywhere in the world, meant to be hungry. In New England, 1816 was nicknamed the "Year without a Summer" or "Eighteen-Hundred-and-Froze-to-Death." Germans called 1817 the "Year of the Beggar." Across the globe, harvests perished in frost and drought or were washed away by flooding rains. Villagers in Vermont survived on porcupine and boiled nettles, while the peasants of Yunnan in China sucked on white clay. Summer tourists traveling in France mistook beggars crowding the roads for armies on the march.

One such group of English tourists, at their lakeside villa near Geneva, passed the cold, crop-killing days by the fire exchanging ghost stories. Mary Shelley's storm-lashed novel *Frankenstein* bears the imprint of the Tambora summer of 1816, and her literary coterie—which included the poets Percy Shelley and Lord Byron—will serve as our occasional tour guides through the suffering worldscape of 1815–18. As one literary historian has observed, "there was never a more documented group of people" than Mary Shelley's circle of friends and lovers. The paper trail of impressions they left of the late 1810s will lead us back again and again to Tambora.[4]

In the early nineteenth century, the overwhelming majority of the world's population were subject (unlike Dr. Frankenstein) to the unforgiving regime of nature: most people depended on subsistence agriculture, living precariously from harvest to harvest. When crops worldwide failed in 1816, and again the next year, starving rural legions from Indonesia to Ireland swarmed out of the countryside to market towns to beg for alms or sell their children in exchange for food. Famine-friendly diseases, cholera and typhus, stalked the globe from India to Italy, while the price of bread and rice, the world's staple foods, skyrocketed with no relief in sight.

Across a European continent ravaged by the Napoleonic Wars, tens of thousands of demobilized veterans found themselves unable to feed their families. They gave vent to their desperation in town-square riots and military-style campaigns of arson, while governments everywhere feared revolution. Human tragedies rarely unfold without blessings to some. During this prolonged global dearth that ended only with the bumper harvests of 1818, farmers in Russia and the American western frontier flourished as never before, selling their grain at stratospheric prices to desperate buyers in the Atlantic trade zone. But for most people, worldwide, this was "the worst of times."

Tambora's aftermath, particularly the "Year without a Summer," 1816, is rich in folklore and continues to be the subject of popular histories. But these accounts are confined to 1816 and to Tambora's impacts in Europe and North America.[5] None has engaged seriously with the robust and ever-increasing scientific literature on Tambora, volcanism, and global climate change. I learned about Mount Tambora in an atmospheric science seminar, not from a history book, and my first exhilarated thought was it was high time historians caught up with the climatologists on Tambora. This book, through all the byways of its creation, remains the product of that initial inspiration. *Tambora* is the first study of this iconic period to marry a volcanological account of the 1815 eruption with both the folklore of the "Year without a Summer" and the full range of biophysical sciences relevant to climate change. It is the first to treat the Tambora event not as the natural disaster of a single year, 1816, but as a three-year episode of drastic climate change whose downstream effects can be traced long into the nineteenth century.

I focus here on the human story of Tambora's aftermath, while my engagement with the detailed discussion of the volcano to be found in recent scientific literature allows me to tell that story on multiple spatial scales, from the molecular to the global. My emphasis, from the point of view of method, is less on nature's impact on history—far less a crude environmental determinism—but on Tambora as a case study in the fragile interdependence of human and natural systems. Put another way, this book considers the disparate human communities of 1815–18—and the climatic zones to which they were adapted—as a single, anthro-ecological world system on which Tambora acted as a massive, traumatic perturbation. After April 1815, many human societies were "changed, changed utterly"—to borrow from the poet

W. B. Yeats—altered, in radical ways, from their pre-eruption state. I have not traveled, in my fieldwork and research, to all continents. In some areas of the globe—notably Africa, Australia, and Latin America—the contemporary data are thin or archives nonexistent. But *Tambora* offers a rich and unique travelogue nonetheless, traversing the hemispheres to trace this epochal eruption's shaping hand on human history.

Across Asia, for example—whose Tambora story has never been told—the volcano's effects were arguably the most devastating of all. A celebrated ancient genre of Chinese verse is called Poetry of the Seven Sorrows. In a Seven Sorrows poem, the poet dramatizes the five bodily senses under assault, overlaid with the twin mental afflictions of injustice and bitterness: seven sorrows in all. The original work in the genre, from the third century, tells of a man forced by civil war from his home, a kind of Chinese Dante. The sorrowful poet, Wang Can, sees lines of corpses from the road and encounters an anguished woman who has abandoned her child in the barren fields. She cannot feed it, but she loiters nearby, listening to its cries. As we will see in chapter 5, the ancient poetic mode of Seven Sorrows enjoyed a renaissance in China in the Tambora period of 1815–18 because it captured so well the human suffering wrought by three successive years of climatic breakdown. A forgotten Chinese poet named Li Yuyang, it turns out, spoke as movingly as anyone for the weather-devastated world of the late 1810s.

The accounts of environmental breakdown and human tragedy left by survivors such as Li Yuyang must stand in for countless histories of individual and community trauma from the Tambora period that are lost forever. In the aftermath of a mega-disaster such as Tambora, the paucity of victim narratives itself tells us something both of the scale of the cataclysm and who bore the brunt of it: the poor and illiterate peasant millions of the early nineteenth-century world. Just as my vision failed me confronted by the dizzying vortices of Tambora's vast caldera, so a complete panoramic view of the human crises it spawned lies out of reach. But with an eye committed to a twenty-first-century way of seeing—to tracing the complex teleconnections between earth, sky, and the fate of human beings—the haywire story of a two-centuries-old global climate crisis may at last be properly told and with it our own fate, in cautionary ways, foretold.

CHAPTER ONE

THE POMPEII OF THE EAST

TIME OF THE ASH RAIN

On April 10, 1815, Napoleon Bonaparte, recently escaped from the island of Elba, was back in Paris and up to his usual tricks. While he charmed one old foe—the liberal journalist Benjamin Constant—into composing a new French constitution guaranteeing democratic rights, he bullied his friend General Davout into raising a half-million-man army. A reenergized Napoleon intended to reclaim full dictatorial powers over France and as much of Europe as possible.

Over in Vienna, on April 10, the aristocratic elite of Europe had cut short their endless round of balls and gourmandizing to hurry up the business of carving up the continent. Every minor prince and dispossessed count of the Old Regime was there to haggle for a fiefdom, while the Great Powers dealt land back and forth like cards in a game of baccarat. "We are completing the sad business of the congress which is the most mean-spirited piece of work ever seen," wrote diplomat Emerich von Dalberg.[1] Meanwhile, the Duke of Wellington, who had rushed from Vienna to organize the allied forces against Bonaparte, had just arrived in Brussels to find it devoid of troops and munitions. With both sides from the twenty-year conflict exhausted and in disarray, all of Europe awaited a messy conflagration, its outcome dubious.

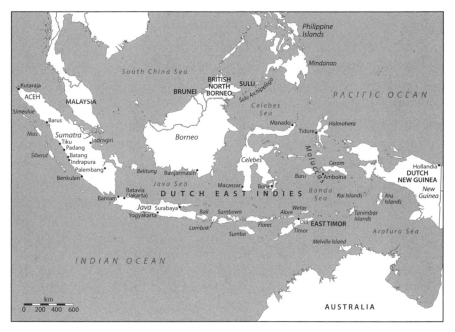

FIGURE 1.1. A map of the nineteenth-century Dutch East Indies. Rubber, spices, rice, tobacco, nickel, and tin were among the commodities sought after by European and Chinese traders. (Based on John Haywood, *Historical Atlas of the 19th Century World* [New York: Barnes & Noble, 2002], 5.22).

Meanwhile, on the far side of the world, on Sumbawa—a remote island outpost of the European war in the blue seas east of Java—the beginning of the dry season in April meant a busy time for the local farmers. In a few weeks the rice would be ready, and the raja of Sanggar, a small kingdom on the northeast coast of the island, would send his people into the fields to harvest. Until then, the men of his village, called Koteh, continued to work in the surrounding forests, chopping down the sandalwood trees vital to shipbuilders in the busy sea lanes of the Dutch East Indies.

In the fields of Sanggar and the neighboring half-dozen island kingdoms, the people cultivated mung beans, corn, and rice, as well as cash crops for the lucrative regional market: coffee, pepper, and cotton. Others collected honey or tiny bird's nests from the seaside cliffs, an aphrodisiac much sought after by wealthy, lovelorn Chinese. In the grassy

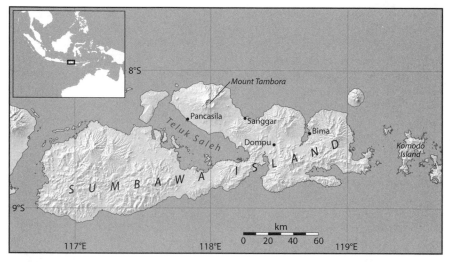

FIGURE 1.2. This map of Sumbawa shows Tambora's dominance of the northeastern Sanggar peninsula, sparsely populated since 1815. The island capital, Bima, lies to the east. Out of picture, to the west, lie the smaller but better-known islands of Lombok and Bali, and beyond these, Java, the principal island in the region. Sumbawa's eastern neighbor is the island of Komodo, home of the famous "dragon" lizard.

fields of the village, meanwhile, Sumbawa's famous horse breeders groomed their stock.[2] For these commodities, the raja and his villagers traded a range of practical and luxury items, including cattle, salt and spices from the islands to the east, bronze bowls from China, and prettily decorated pots from what are now Cambodia and Vietnam.[3]

Sumbawa had been settled by people from the neighboring larger islands of Java, Celebes, and Flores about four hundred years earlier. These pioneers converted large stretches of the densely wooded landscape to rice paddies and grasslands for grazing cattle and horses. Pre-eruption Sumbawa boasted a great diversity of ethnicities and languages. The Sanggarese, for example, on the northeastern peninsula, looked nothing like their compatriots on the western side of Sumbawa; nor could they understand each other's speech. Of the mother islands, Celebes had maintained the strongest influence on Sumbawa as a kind of vassal state, its powerful capital Macassar exacting crushing taxes upon the Sumbawans. Then, in the seventeenth century, the Dutch arrived to stake their claim upon the region. It was the good fortune of the Sumbawans that the Dutch took little interest in their island while at the same time curbing the power of Macassar. The early nineteenth century, then, saw Sumbawa in a better position than it had ever known: economically integrated in the region but with a degree of political independence.

Even in the midst of this prosperity, however, the raja of Sanggar could never relax. Now that sailing conditions had improved after the abating of the rains, he kept a wary eye out for pirates from the Sulu islands to the north, who preyed upon coastal villages looking for human prizes for the slave market.[4] With sleek sailboats carrying up to a hundred armed men, and a gift for surprise attack, the pirates were a terrible sight for the people of Sanggar. Given sufficient warning, the young men might escape deep into the forest. But everyone was vulnerable, and the raja could not have thought of his own children without anxiety. Once taken by pirates, their inherited privileges—and their happy village life on Sumbawa—would be lost forever. But slavery was a fact of life, nevertheless. While resources were abundant in the islands, labor was not. Human beings were thus the most valuable of commodities, and the oceangoing traffic in flesh was cruel and unrelenting. Between the 1770s and 1840s, several hundred thousand people passed through the slave markets of the East Indies, the largest slave system outside the Atlantic zone.[5]

Another source of anxiety lay closer at hand: the magnificent mountain Tambora, the tallest peak in an archipelago rich in cloudy, volcanic summits. The broad, forested slopes of Tambora dominated the Sanggar peninsula, and its distinctive twin peaks served as a major navigation point for shipping—and pirates. The long-dormant Tambora had for some years past begun to rumble periodically, sending forth dark clouds from its airy summit. A British ship captained by the diplomat and naturalist John Crawfurd sailed near the belching mountain in 1814:

> At a distance, the clouds of ashes which it threw out blackened one side of the horizon in such a manner as to convey the appearance of a threatening tropical squall. . . . As we approached, the real nature of the phenomenon became apparent, and ashes even fell on the deck.[6]

Local opinion varied as to the cause of the mountain's groggy awakening. Some thought it the celebration of a marriage among the gods,
while others viewed it more darkly. The rumblings signified anger, they
said. In a notorious incident, a Sumbawan chief had murdered a Mus-
lim pilgrim. Another legend still popular in Sumbawa tells of a visiting
"shaykh," a holy man, who was outraged to find dogs loose in the local
mosque. When the offended locals served him canine meat in revenge,
the shaykh discovered the trick and began to pray. In an instant he vanished, the butchered dog reappeared in living form, and the volcano
began to bellow.[7] Still others believed the gods were angry that the people had allowed foreign white men with their ships and guns to enslave
them on plantations on nearby Java and Macassar.[8]

The raja took all these opinions personally. Throughout the East Indies, volcanism served as a symbol of political power. Sultans, for example, represented themselves as offspring of the mountain god, Siva.[9]
Volcanic eruptions were accordingly viewed as mirrors of human affairs, as punishment for the poor administration of their rulers. Tambora's rumblings were bad news for the raja; they unnerved his people
and undermined his legitimacy in their eyes.

On the evening of April 5, 1815, at about the time his servants would
have been clearing the dinner dishes, the raja heard an enormous thun-
derclap.[10] Perhaps his first panicked thought was that the beach lookout
had fallen asleep and allowed a pirate ship to creep into shore and fire
its cannon. But everyone was instead staring up at Mount Tambora.
A skyward jet of flame burst from the summit, lighting up the darkness and rocking the earth beneath their feet. The noise was incredible,
painful.[11]

Huge plumes of flame issued from the mountain for three hours,
until the dark mist of ash became confused with the natural darkness,
seeming to announce the end of the world. Then, as suddenly as it had
begun, the column of fire collapsed, the earth stopped shaking, and the
bone-jarring roars faded. Over the next few days, Tambora continued
to bellow occasionally, while ash drifted down from the sky. But for
the raja the emergency appeared over. His first concern was for the imminent rice harvest. The villagers toiled night and day in the fields to

clean the thick film of gray, sandy dust from the rice plants—a messy business.

Meanwhile to the southeast in the capital Bima, colonial administrators were sufficiently alarmed by the events of April 5 to send an official, named Israel, to investigate the emergency situation on the Tambora peninsula. We don't know if he stopped to discuss the situation with the raja of Sanggar, but by April 10 the unlucky man's bureaucratic zeal had led him to the very slopes of Tambora. There, in the dense tropical forest, at about 7:00 PM, he became one of the first victims of the most powerful volcanic eruption in recorded human history.

Within hours, the village of Koteh, along with all other villages on the Sanggar peninsula, ceased to exist entirely, a victim of Tambora's spasm of self-destruction. This time three distinct columns of fire burst in a cacophonous roar from the summit to the west, blanketing the stars and uniting in a ball of swirling flame at a height greater than the eruption of five days before. The mountain itself began to glow as streams of boiling liquefied rock coursed down its slopes. At 8:00 PM, the terrifying conditions across Sanggar grew worse still, as a hail of pumice stones descended, some "as large as two fists," mixed with a downpour of hot rain and ash. A decade after the event, a native poet from Bima described the horrific scene:

> The mountain reverberated around us
> As torrents of water mixed with ash fell from the sky.
> Children screamed and wept, and their mothers, too,
> Believing the world had been turned to burning ash.[12]

On the north and western slopes of the volcano, whole villages, totaling perhaps ten thousand people, had already been consumed within a vortical hell of flames, ash, boiling magma, and hurricane-strength winds. In 2004, an archaeological team from the University of Rhode Island uncovered the first remains of a village buried by the eruption: a single house under three meters of volcanic pumice and ash.[13] Inside the walled remains, they found two carbonized bodies, perhaps a married couple. The woman, her bones turned to charcoal by the heat, lay on her back, arms extended, holding a long knife. Her sarong, also carbonized, still hung across her shoulder. She had been interrupted by fiery death while at a mundane domestic task—preparing the evening meal—much like the petrified figures of women, children, and household pets at Pompeii already famous in the early nineteenth century. The Tambora woman's charcoal state, however, is evidence of immolation at far higher temperatures than those generated by Vesuvius in AD 79.

Back on the mountain's eastern flank, the rain of volcanic rocks gave way to ashfall, but there was to be no relief for the surviving villagers. The spectacular, jet-like "plinian" eruption (named for Pliny the Younger, who left a famous account of Vesuvius's vertical column of fire) continued unabated, while glowing, fast-moving currents of rock and magma, called "pyroclastic streams," generated enormous phoenix clouds of choking dust. As these burning magmatic rivers poured into the cool sea, secondary explosions redoubled the aerial ash cloud created by the original plinian jet. An enormous curtain of steam and ash clouds rose and encircled the peninsula, creating, for those trapped inside it, a short-term microclimate of pure horror.

First, a "violent whirlwind" struck Koteh, blowing away roofs. As it gained in strength, the volcanic hurricane uprooted large trees and launched them like burning javelins into the sea. Horses, cattle, and people alike flew upward in the fiery wind. What survivors remained then faced another deadly element: giant waves from the sea. A British ship cruising offshore in the Flores strait, coated with ash and bombarded by volcanic rocks, watched stupefied as a twelve-foot-high tsunami washed away the rice fields and huts along the Sanggar coast. Then, as if the combined cataclysms of air and sea weren't enough, the land itself began to sink as the collapse of Tambora's cone produced waves of subsidence across the plain.

It is hard to believe anyone could survive such a hellhole of destruction, but the raja of Sanggar and members of his family somehow did get away, along with a few dozen members of his village. Perhaps the raja exerted his royal privileges in claiming the best horses from the stables and set out early enough on the terrible evening of April 10 to

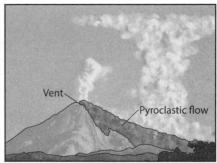

FIGURE 1.3. After the initial plinian jet of a Tambora-style eruption, the volcano's vent collapses, and pyroclastic flows of magma issue down the subsiding mountain slopes. In the case of an island eruption such as Tambora's, these boiling streams generate enormous secondary ash clouds on flowing into the cooler sea waters. Thus both the original plinian event and the subsequent phoenix clouds are capable of injecting volcanic matter into the stratosphere, as happened in 1815.

escape the eruption's reach, following an inland southerly course away from the sea, whose interaction with the pyroclastic streams created the deadly whirlwind and tsunami. Keeping to the narrow route between Koteh and Dompu—the sole band of the peninsula spared from lava inundation—their unimaginable flight involved five-meter-high molten rivers spitting and smoking on either side, their escape like a latter-day miracle of the Red Sea. The raja of Sanggar and his band of survivors owed what life remained to them to both the topography of Tambora, which directed the magmatic flow of the April 10 eruption more to the northwest and south, and the trade winds, which blew the volcanic ash in a westward direction toward the islands of Bali and Java.

On the sunless days following the cataclysm, corpses lay unburied all along the roads on the inhabited eastern side of the island between Dompu and Bima. Villages stood deserted, their surviving inhabitants having scattered in search of food. With forests and rice paddies destroyed, and the island's wells poisoned by volcanic ash, some forty thousand islanders would perish from sickness and starvation in the ensuing weeks, bringing the estimated death toll from the eruption to over one hundred thousand, the largest in history.[14] Even the wealthy raja of the now vanished kingdom of Sanggar could not save his beloved daughter, weakened by terror and an unrelenting diarrhea brought on by ash-poisoned water.

One day, after many weeks, the raja heard that Englishmen had come to the island with a ship full of rice. He hurried to Dompu, where he used his title to gain an audience with the English chief, Lieutenant Owen Phillips of the navy. Desperate and grief-stricken as he was, the raja could not but be wary on being led alone into the presence of the English governor's envoy. In the local zoonomia of races, the Dutch overlords were horse leeches draining the native people of their blood, while the Sumbawans themselves resembled the buffalo: solid, long-suffering beasts of burden. But these new conquerors, the British, appeared like red-faced tigers, down to the animal skins the officers wore to decorate themselves: dazzling but deadly.[15]

Having survived the terrible eruption of Tambora, however, the raja of Sanggar possessed courage and wit enough for Lieutenant Phillips of the Royal Navy. He gave Phillips his description of what had happened on the Sanggar peninsula on April 10, 1815—the sole existing eyewitness account of Tambora's mighty explosion. He told his story well enough that the English officer awarded him several tons of rice to feed his people and quoted him at length in dispatches. The raja thanked the Englishman with emotion but left quickly, no doubt with a mind still focused on survival and little recognizing his service to history.

THE GOLDEN KINGDOM OF TAMBORA

Amazingly, Tambora's twin plinian explosions accounted for only about 4% of the volcano's eruptive production. While the skyward eruptions lasted only about three hours each, the boiling cascade of pyroclastic streams down Tambora's slopes continued a full day. Hot magma gushed from Tambora's collapsing chamber down to the peninsula, while columns of ash, gas, and rock rose and fell, feeding the flow. The fiery flood that consumed the Sanggar peninsula, traveling up to thirty kilometers at great speeds, ultimately extended over a 560-kilometer area, one of the greatest pyroclastic events in the historical record. Within a few short hours, it buried human civilization in northeast Sumbawa under a smoking meter-high layer of ignimbrite.

The rivers of volcanic matter that plowed into the sea redrew the map of Sanggar. Forests were incinerated along with villages, and kilometers of coastline added to the peninsula, like a giant volcanic landfill. Once Tambora had disgorged itself of its subterranean sea of magma, its mountain shell imploded. With so vast a volume of interior matter expelled, catastrophic subsidence was inevitable, and sometime on April 11 or 12 Tambora sank into itself, forming a six-kilometer-wide caldera where once its lofty peak had been. This giant volcanic sinkhole is among the largest to have formed on Earth since the retreat of the last glacial period around twelve thousand years ago and is comparable in size to the volcanic Crater Lake in Oregon, formed seven thousand years ago. In all, Tambora lost a kilometer and a half in height during its week-long rage of self-destruction. Where once it rose to a classical conical peak, it now rested from its labors like a long, recumbent giant on an expanded lava bed, denuded of life.

Tambora's cacophony of explosions on April 10, 1815, could be heard hundreds of miles away. All across the region, government ships put to sea in search of imaginary pirates and invading navies.[16] In the seas to the north off Macassar, the captain of the East India Company vessel *Benares* gave a vivid account of conditions in the region on April 11:

> The ashes now began to fall in showers, and the appearance altogether was truly awful and alarming. By noon, the light that had remained in the eastern part of the horizon disappeared, and complete darkness had covered the face of day.... The darkness was so profound throughout the remainder of the day, that I never saw anything equal to it in the darkest night; it was impossible to see your hand when held up close to the eye.... The appearance of the ship, when daylight returned, was most extraordinary; the masts, rigging, decks, and every part being covered with falling matter; it had the appearance of a calcined pumice stone, nearly the colour of wood ashes; it lay in heaps of a foot in depth in many parts of the deck, and I am convinced several tons weight were thrown overboard.[17]

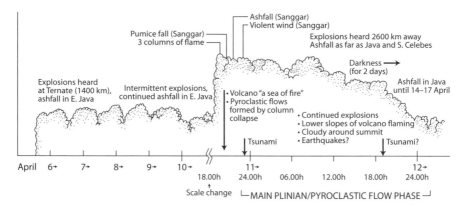

FIGURE 1.4. Timeline of Tambora's eruption, based on eyewitness accounts and subsequent geological analysis of the site. (Adapted from Stephen Self et al., "Volcanological Study of the Great Tambora Eruption of 1815," *Geology* 12 [November 1984]: 660).

For years after 1815, ships encountered vast islands of pumice stone as far away as the Indian Ocean, thousands of kilometers to the west. These great pumice pontoons were littered with burnt slivers of trees, the carbonized residue of Sumbawa's once dense and valuable forests.

During the eruption, on Borneo to the north, where the earth shook amid a great roar, the indigenous people believed the sky was falling, while on the eastern coast of Java, the birds kept a stunned silence until 11:00 the next morning. Visibility shrank to a few feet, so thick was the fallout. Driven west by the beginning monsoon, Tambora's ash cloud consumed Sumbawa and Lombok before descending on Bali, covering the island in ash a half-meter deep.[18] The same gigantic cloud "dreary and terrific"—Tambora's pyroclastic plumes drifting westward—could then be seen approaching the Javan shore from the direction of Bali, and the air grew very cold.[19] Across a 600-kilometer radius, darkness descended for two days, while Tambora's ash cloud expanded to cover a region nearly the size of the continental United States. The entire Southeast Asian region was blanketed in volcanic debris for a week. Day after dark day, British officials conducted business by candlelight, as the death toll mounted.

When a semblance of sunlight returned at last sometime on April 13, the scattered bands of survivors on the Sanggar peninsula found them-

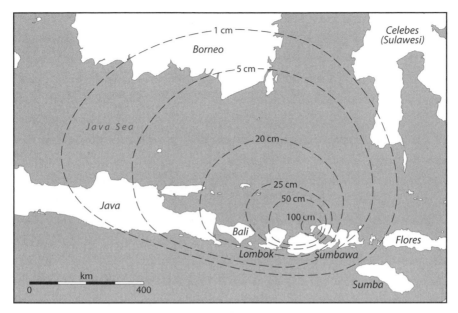

FIGURE 1.5. Map showing the density of ashfall issuing from Tambora's phoenix clouds (the plinian explosion, because vertical, produced ashfall across a smaller area). Prevailing trade winds drove the ash clouds north and west as far as Celebes (Sulawesi) and Borneo, 1,300 kilometers away. The explosions on April 10, 1815, could be heard twice as far away. (Adapted from Stephen Self et al., "Volcanological Study of the Great Tambora Eruption of 1815," *Geology* 12 [November 1984]: 661).

selves in an unrecognizable, barren landscape. In every direction, ash meters thick had buried their island home as they had known it. On the western slope of the mountain, where the kingdom of "Tambora" itself was submerged, an entire ethnic group disappeared, and with it their language, the easternmost Austro-Asiatic tongue. On nearby islands, conditions were almost as dire. Reports later surfaced of starvation and rat plagues on Lombok, while thousands of Balinese attempted to sell themselves or their children for handfuls of rice.[20]

A worse fate met those *unable* to resort to the slave market. As a response to abolitionist legislation recently enacted by the British Parliament, the governor of Java, Stamford Raffles, had outlawed the slave trade in the capital Batavia (now Djakarta), unwittingly eliminating the only social safety net his subjects knew.[21] One wonders if Raffles ever understood the unforeseen consequences of that progressive policy, whether reports reached him of child corpses lining the beaches on Bali, killed by parents unable to sell them for food and presumably unwilling to watch them suffer the slow starvation they themselves faced.[22]

Months after the eruption, the atmosphere remained heavy with dust—the sun a blur. Drinking water contaminated by fluorine-rich ash spread disease and with 95% of the rice crop in the field at the time of the eruption, the threat of starvation was immediate and universal. In their desperation for food, islanders were reduced to eating dry leaves and their much-valued horseflesh. By the time the acute starvation crisis was over, Sumbawa had lost half its population to famine and disease, while most of the rest had fled to other islands.

As late as 1831, sixteen years after the eruption, northeast Sumbawa still resembled a war zone, as if the disaster had just occurred. A Dutch official sailing along the coast observed through his eyeglass "a horrendous scene of devastation . . . in its fury, the eruption . . . has spared, of the inhabitants, not a single person, of the fauna, not a worm, of the flora, not a blade of grass."[23] A recent tree-ring study has shown that the entire Java region suffered a drastic period of cold, drought conditions in the aftermath of the eruption.[24] On account of the wholesale deforestation of the island, the Sumbawan micro-climate changed radically, becoming far drier. The longer-term social impacts were just as dismal. A half century later, a visitor found Sumbawa populated mostly by slaves descended from survivors of Tambora who had sold themselves into bondage. As a result of these flow-on disasters, the Sanggar peninsula has never been fully repopulated, and Sumbawa Island as a whole never recovered.

History records that four years after Tambora's eruption—his colonizing ambitions for Java dashed and the region's climate returned to normal—Stamford Raffles founded a new colony at Singapore. With that one stroke, he transformed the balance of trade and power in Britain's favor in the East Indies. But on Sumbawa, where the light of Western historiography barely shines, the local people still refer to the apocalyptic eruption of 1815 as the moment their world changed forever. Just as the Holocaust is *Shoah* to the Jews, so the Tambora disaster bears its own sanctified name for the Sumbawans: *zaman hujan au* (time of the ash rain).[25] As I traveled across the island two centuries later, over barely passable roads and through meager townships without clean water or sanitation, it was evident that Sumbawa still lived in the shadow of Tambora. The year following my visit, at least twenty Sumbawan children were reported as having died from malnutrition.[26] On the long, bumpy drive back from Tambora to Bima, my guides—from the more prosperous neighboring island of Lombok—made endless fun of the backwardness and poverty of the locals.

But was it always so? International forestry companies dominate the Sumbawan economy today. This, in addition to rampant illegal logging, is gradually repeating the process of deforestation wrought by Tambora in a single day two hundred years ago. In 1980, a forestry company came upon the remains of a "lost kingdom" of Tambora on the western slopes of the mountain. Beneath a thin humus layer of new-growth forest on Tambora sits a meter of compacted ignimbrite deposited by the 1815 eruption. Beneath that, loggers uncovered a cache of Chinese-patterned pottery shards and burned human bone fragments. Locals soon showed up with brass pots, jewelry, and eighteenth-century Dutch coins they said also belonged to the site—a blurry snapshot of the unrecovered Pompeii of the East.[27]

Sumbawans still fear their volcano and tell visitors that it will surely erupt again soon. For that reason they inhabit villages at a respectful distance from Tambora and maintain its legend through stories. They talk of the island's great wealth before the *zaman hujan au*, when a rich king ruled the mountain from a palace of gold.[28] The ghost of the Tambora king haunts the mountain forests, where his treasure is buried. He is of a wily and embittered disposition, and addicted to spells. If one must go to Tambora, Sumbawans say, be wary of this ghost king. No bad language; no speaking ill of others. And if a young man brings his girlfriend into the Tamboran forest, he must not think of making love to her there. For once the ghost king's vengeful attention is roused, there can be no escape. At first, it will seem like a wonderful dream. The ghost king will reveal his lost kingdom with its palace of gold to your disbelieving eyes. Out of the jungle, trees bearing delicious fruit will appear in dazzling colors, promising bliss. And to complete the enchantment,

FIGURE 1.6. Pottery and other household items excavated from a buried village on the lower slopes of Tambora in 2004. They offer evidence of a prosperous island community well integrated in the regional trade zone of the East Indies and, by extension, linked to the hemispheric economies of China and Europe. (© University of Rhode Island/Michael Salerno).

the ghost king will send his own daughter, another phantom of 1815, to lure you ever deeper into the forest with her trilling laughter, until you can never find your way home. Many young Sumbawan men have been lost this way, they say.

We climbed the jungle slopes of Tambora late in the rainy season. Tormented by leeches and the razor-sharp leaves of a fern called *srra*, we came upon a trio of young Sumbawans hunting wild pig in the jungle. They were smoking cigarettes under cover of a small lean-to after a fruitless morning's hunt. From a tinny cassette player in the dirt, a young woman sang songs of love in Arabic, accompanied by the jungle's hum. The young hunters, good Sumbawans all, listened to her gentle pleadings in silence, unmoved. When we had rested—and picked each other clean of leeches—the men showed no inclination to join us on our climb. Only the occasional volcanologist or tourist now visits Tambora's sprawling cratered peak, while its peninsular surrounds are mostly populated by recent immigrants from other islands. Deserted but for these, the great headless mountain Tambora lives on as the lotus land of a traumatized Sumbawan imagination.

THE PHILOSOPHER KING OF JAVA

Arab traders in quest of spices were the first to arrive in the East Indian seas surrounding Mount Tambora. Their influence endures in the prevalence of Islam in the region. The Portuguese followed in the sixteenth century; then, as their maritime power waned, the Dutch asserted themselves as colonial masters. The Dutch East India Company established cash crop production—pepper, coffee, sugar—across the Java archipelago, while delegating daily management of the estates to Chinese middlemen who brutalized their laborers to meet production targets. The Dutch brought nothing to make or sell in the East Indies; instead they exploited it as a vast farm belt, its riches to be sold on the European market at fabulously inflated prices. The well-fed Dutch burghers who gaze contentedly from seventeenth-century portraits by Rembrandt and Frans Hals were the beneficiaries of this cascade of wealth emanating from the East Indies.

It is thus an historical fluke that places the British in control of Sumbawa at the time of Tambora's sudden explosion in 1815. Britain's rule of Java and its surrounding islands marks a brief interregnum in the centuries-long influence of Arab, Portuguese, and Dutch merchants. Nevertheless, Stamford Raffles's imprint on Java was significant. After asserting control of the island by military force, Raffles handpicked local sultans to rule over the principalities of Java and appropriated vast acreages of land. Thus far, a familiar tale of European conquest. Raffles differed radically from his Dutch predecessors in one respect, however: he had a profound interest in East Indies culture.

Two centuries of Dutch rule produced scarcely a single paper on any aspect of Javan history, customs, or language. In the brief five years of Raffles's tenure in Java, by contrast, he fashioned himself as a kind

FIGURE 1.7. Portrait of Sir Stamford Raffles, by George Francis Joseph (1817). The land-scape background suggests the fertile "Eden" of the East Indies that was Raffles's po-litical dominion, while the rich assortment of papers and Asian artifacts represents his scholarly investment in the *History of Java*, published the year of the portrait. The Hindu character of Raffles's art collection is deliberate: Raffles's *History* goes to great pains to elevate Java's "native" Hindu traditions, more congenial to English colonizers, above the more recent introduction of Islamic culture. (© National Portrait Gallery, London.)

of philosopher king or, more accurately perhaps, as an anthropologist with a bottomless research account, a well-armed regiment at his disposal, and no professional protocols to bother him. He learned Javanese, in addition to his already fluent Malay, bought whatever historical manuscripts he could find, and employed fleets of local copyists to manufacture a library of Javanese source material. He sent an army of assistants into the field to collect flora, fauna, and geological specimens of all kinds, as well as artists to make drawings. Once news of the English governor's mania for collecting spread, the prince of a nearby island sent him an orangutan, which Raffles dressed up like a man in coat and trousers and hat.[29] Raffles would go on to become the founder and first president of the London Zoo.

The governor presided over this extensive menagerie and research factory as its prose synthesizer, a kind of mythmaker-in-chief of Javanese history. After entertaining his European colleagues over dinner, Raffles would repair to his study, where pen, ink, and paper lay ready for him, with two large candles lit. After pacing up and down for a time, he would lie on the table with his eyes closed as if asleep, then spring up and write furiously past midnight. In the morning, he read over what he had written, saving three sheets out of ten and tearing up the rest.

The result of this frenzy of scholarly activity has been called the first "classic of South East Asian historiography."[30] The two volumes of Raffles's *History of Java* (1817) aim at a comprehensive account of Javan culture in the Western enlightenment style, organized under rubrics of geology and geography, agriculture and manufacturing, language and customs, history and government. For all its apparatus of scholarly objectivity, however, the *History of Java* promotes a specific political agenda at every turn. Interwoven with botanical descriptions and historical accounts are vivid threads of travelogue propaganda, anti-Dutch polemic, and reformist colonial policy, all designed to promote the cause of continued British administration of the Javan archipelago, including Sumbawa.

To his East India Company superiors, Raffles paints a lyrical picture of the Java region as an Eden of agricultural possibility, ripe for European development:

Nothing can be conceived more beautiful to the eye, or more gratifying to the imagination, than the prospect of the rich variety of hill and dale, of rich plantations and fruit trees or forests, of natural streams and artificial currents . . . it is difficult to say whether the admirer of landscape, or the cultivator of the ground, will be most gratified by the view. The whole country, as seen from mountains of considerable elevation, appears a rich, diversified, and well-watered garden.[31]

The economic theorist Adam Smith, whom Raffles frequently quotes, abominated the Dutch colonial monopoly. He advised administrators of the rising British empire to abandon monopolist economics in favor of a free-trade system, so as "to open the most extensive market for the produce of his country, to allow the most perfect freedom of commerce . . . to increase as much as possible the number and competition of buyers."[32] Java and its surrounding islands were thus not a lost cause, Raffles argued, but a natural laboratory for free-market economics and a golden opportunity for a progressive colonial power to enrich both itself and its subjects.[33]

All good in theory, of course, except for the small matter of volcanoes. Raffles's challenge in 1815, as a self-styled imperial visionary, was to adapt his hothouse modern ideas of free trade and political liberty to a wholly different cultural ecology: one in which everything—from the fertile soil beneath his feet, to the omnipresent mountains set against the sky, to the periodic matter of that sky itself—was volcanic, and whose inhabitants measured history by the remembered cataclysm of eruptions.

In Batavia, on the morning of April 11, 1815, Raffles woke late in darkness, having dreamed his nightly fill of British dominion over the East Indies. By the time he was wading through knee-high ash in his vice-regal garden, ten thousand of his Sumbawan subjects were already dead. His initial response to the catastrophe was in character as both a modern bureaucrat and a scholar: he demanded full written reports of the event from his regional subordinates. But the full extent of the devastation appears to have dawned fatally slowly on the British governor. It was not until August, on hearing reports of famine on Sumbawa, that

FIGURE 1.8. An idealized Javan landscape from the early nineteenth century paints an idyllic scene; but it also shows the proximity of village life to the volcanic mountains that run east-west across the East Indies archipelago. (Lady Sophia Raffles, *Memoirs of Sir Stamford Raffles* [London: John Murray, 1830], 149.)

he sent a ship laden with rice as a form of disaster aid, with Lieutenant Owen Phillips in charge of relief operations. Raffles takes pride in this act in his *History*, though by our modern reckoning his humanitarian gesture was pitifully inadequate: a mere few hundred tons of rice, capable of feeding perhaps twenty thousand survivors on Sumbawa for a week.

Given the cataclysmic scale of Tambora, it is also accorded strangely minimal space in Raffles's *History of Java*. His narrative of the eruption is not to be found in the long chapters on Javan history or even in the generous section on volcanoes. Rather, it is squeezed into a lengthy footnote between essays on Java's "Mineralogical Constitution" and its "Seasons and Climate" as an episode that "may not be uninteresting" to his readers. Raffles thus stands as the first in a long line of Western historians to miss, or in his case willfully deny, Tambora's impact. His reticence on Tambora is easily explained. Raffles represents Tambora as no more than a natural wonder, a volcanic sound-and-light show, because his argument for a British Java could only be undermined by its proximity to giant volcanoes capable of plunging the entire regional economy into ruin in the course of a few hours.

Tambora thus begins as a natural disaster story—a Pompeii of the East or a hundredfold Hurricane Katrina—whose fate was to remain largely unwritten. Raffles had assembled the beginnings of a "Temboran" dictionary shortly before the eruption, which he included as a kind of epitaph in an appendix to his *History of Java*. But like Raffles's dictionary, the Tambora story has only ever been told in notes and sketches, with enormous gaps left unfilled. In place of a world-historical narrative, the colossal eruption on Sumbawa Island in 1815 has survived—in faraway countries and other languages—only as a weather folktale: the fabled "Year without a Summer." As the following chapters will show, however, Tambora's world-altering reach requires an epic telling far beyond the frosted memories of a single long-ago summer.

CHAPTER TWO

THE LITTLE (VOLCANIC) ICE AGE

THE VOLCANO LOVERS

On a clear winter's day in early 1819, Mary and Percy Shelley visited the ruins of Pompeii, outside Naples. "I stood within the city disinterred," as Percy remembered it.[1] The excavation of Pompeii, a half century before, had brought volcanism alive to the imaginations of Europeans. The uncarthed city presented a stunning image of human calamity in the face of a major eruption. The Shelleys wandered among the grand theaters, villas, and neatly designed streets of an advanced society that vanished overnight in AD 79. That Vesuvius had recently awoken from a period of dormancy to offer belching reminders of its power perfected the scene for the Romantic tourist. Shelley observed Vesuvius "rolling forth volumes of thick white smoke," prompting his active imagination to conjur the terrifying fate of the inhabitants of Pompeii. Trying his hand at popular volcanology in a letter to his friend Thomas Love Peacock, Shelley theorized that "the mode of destruction is this. First an earthquake shattered it & unroofed almost all its temples & split its columns, then a rain of light small pumice stones fell, then torrents of boiling water mixed with ashes."[2]

Volcanic eruptions were all the rage in the early nineteenth century. The travel writings of pioneer earth scientist Alexander von Humboldt—published the year of Tambora's eruption—offered English readers breathless accounts of the majestic Cotopaxi in the Andes and the smoking volcanic peak of Tenerife. The Scottish naturalist George Steuart Mackenzie, meanwhile, had published his own account of the volcanoes of Iceland, where he came upon the devastated landscape marking a recent eruption:

> The scene now before us was exceedingly dismal. The surface was covered with black cinders; and the various shallows enclosed by high cliffs and rugged peaks destitute of every sign of vegetation, and rendered more gloomy by floating mist, and a perfect stillness, contributed to excite strong feelings of horror.[3]

Because volcanoes stood for so tantalizing a cocktail of emotions—a mix of horror and pleasure, shaken and stirred—they became staple images of poetic description in the Romantic age. Thanks to the vivid first-hand accounts of Sir William Hamilton—amateur volcanologist and British envoy to Naples in the 1760s and 1770s—the ascent of Vesuvius became a highlight of the Grand Tour. The celebrity scientist Humphry Davy climbed the boiling summit fourteen times in 1819–20, taking samples of lava for chemical analysis: "its surface appear[ed] in violent agitation, large bubbles rising, which, in bursting, produced a white smoke."[4] Only months before, the Shelleys had prepared for their ascent by reading Madame de Staël's popular Vesuvian novel *Corinne* (1806), whose more magmatic style may be judged by the following excerpt: "The river of fire flowing from Vesuvius was revealed by the darkness of night, and it seized and bound the imagination of Oswald. Corinne used this impression to turn him from recollections that tormented him, and she hastened to lead him with her away from the inflamed lava."[5] Meanwhile, half a world away, in the Louisiana Territory, an old Mississippi River flatboatman was heard to blame the 1811–12 New Madrid earthquakes, the worst in American history, on "old Vesuvius himself."[6] The power of Vesuvius, as cultural icon, was truly hemispheric.

Back in England, entertainment entrepreneurs offered a variety of volcanic productions for those unable to afford a sojourn to Italy. In the popular pleasure gardens of London sideshow engineers repackaged

FIGURE 2.1. J.M.W. Turner, *Mt. Vesuvius in Eruption* (1812). Turner had never visited the Bay of Naples when he painted this dramatic rendering of Vesuvius. The image thus represents less a physical landscape than a product of Romantic popular culture—an imagined pastiche of verbal descriptions, other paintings, and sideshow volcanic spectacles in London. (Yale Center for British Art.)

their pyrotechnic displays as volcanic paroxysm, treating patrons to nightly effusions of smoke, cacophonic rumbles, and fiery lights jetting from giant plaster Vesuvian cones. "The Eruption of Vesuvius Vomiting Forth Torrents of Fire!" promised a typical newspaper advertisement. Surrey Gardens, which boasted its own lake, could re-create the entire Bay of Naples in its Vesuvius show. The fireworks' reflections in the water enhanced the wow-factor of the spectacle, putting the gardens' managers one step ahead of the competition.[7]

With the violent upheavals of the French Revolution, a new layer of symbolism became adhered to volcanic spectacle.[8] On October 17, 1793, at the height of Robespierre's Reign of Terror, Queen Marie Antoinette faced the guillotine in front of a heaving Jacobin mob in the newly christened Place de la Révolution in Paris. The following day, the nearby Theatre de la République premiered a play titled

Le jugement dernier des rois (*Day of Judgement for the Kings*), authored by hard-core revolutionary journalist Sylvain Maréchal. In the play, a wretched assembly of deposed European monarchs, together with the pope, is deposited on a tropical island under the loom of a belching volcano. After some valedictory abuse from their Jacobin captors, they are left to be consumed in a river of molten lava symbolizing the insurgent wrath of the French people. One Parisian newspaper encouraged the public to attend, promising "you will see all the tyrants of Europe obliged to devour one another and be swallowed up, at the end, by a volcano. There's a show made for republican eyes!" Audiences greeted the tyrannicidal scene with laughter and applause. Revolutionary officials were so impressed with Maréchal's play they distributed thousands of copies to the troops and requisitioned precious gunpowder to keep the volcanic fires of revolution burning through an extended season.[9]

From an historical point of view, the iconic Vesuvius thus represents more than a mere "special effect" of the Revolutionary period, 1789–1830. Volcanism loomed large in the early nineteenth-century European imagination as a readymade symbol for the wave-upon-wave of social crises ordinary people experienced first as an upsurge of violence near at hand: in dead bodies on the street, soldiers pillaging farms, or smashed windows in the market square. The destructive spasms of the erupting volcano seemed the most apt image for the unprecedented bloodletting and upheaval that swept civilian Europe in the decades after 1790.

THE 1810s: COLDEST OF THE COLD

While Vesuvius stood tall in the European imagination, it nevertheless had no impact whatever on global climate. Its intermittent eruptions were orders of magnitude too small. How ironic, then, that Europeans of the late 1810s, distracted by their Vesuvian sideshow, remained oblivious to the planet-wide volcanic crisis through which they lived. Given the state of science and communications, however—no steamships or telegraph, let alone satellites—it is no surprise they failed to "teleconnect" the dots.

Tambora's violent impact on global weather patterns was due, in part, to the already unstable conditions prevailing at the time of its eruption. A major tropical volcano had blown up six years prior, in 1809. This cooling event, hugely amplified by the sublime Tambora in 1815, ensured extreme volcanic weather across the entire decade. We have seen that the historical record surrounding Tambora's eruption is slim: a scattering of reports from Raffles's lieutenants and translated stanzas from a Sumbawan poem are all we have. So obscured from human view is Tambora's immediate predecessor, however, that even its location remains a mystery. The historical fact of the eruption scientists refer to as the 1809 Unknown was only established through the technology of ice core analysis developed in the 1960s.[10]

Ice cores are cylinder-shaped records of annual snowfall, dating back thousands of years, extracted from glaciers and ice sheets at the planet's peaks and poles. These glittering sheaths of ice are among the most beautiful natural artifacts ever to be subject to scientific scrutiny—and the most important. Even given the bone-chilling requirements of their extraction, it is difficult not to envy the scientist whose task it is to unlock the climatic secrets of a giant ice core, suffused with tiny bubbles and dancing with blue light. While the Shelleys and their peers contemplated the work of centuries among the ruins of Pompeii, modern paleoclimatologists see millennia stretched out before them in perfect Apollonian torsos of glacial ice.

A paper published in 1991, based on ice core evidence from both poles, announced the surprising presence of a rich sulfate deposit corresponding to the 1810 and 1811 snowfalls, indicative of a major tropical eruption comparable in size to the famous Krakatau eruption of 1883 (a magnitude approximately half that of Tambora's).[11] A subsequent ice core study, based on sulfate deposits across the Greenland ice sheet, rated the 1809 Unknown the *third* largest eruption since the early 1400s, behind only Tambora and the eruption of Mount Kuwae on Vanuatu in 1452.[12] Certainly, many parts of the globe felt its cooling impact the following year. In parts of England, the first days of May 1810 were the coldest in living memory, bitten with late frosts, while the lowlands hills of Scotland remained eerily white through the spring. In Manchester,

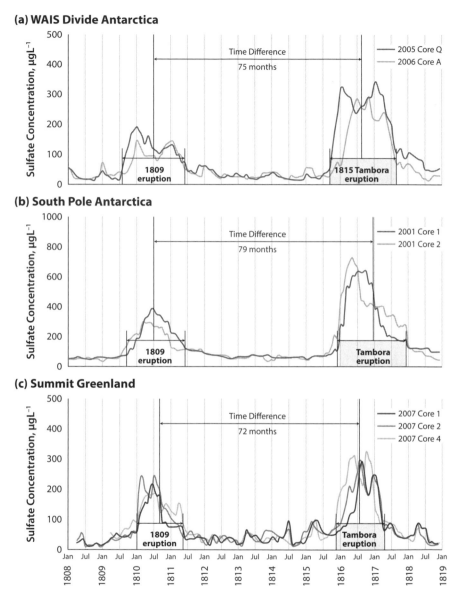

FIGURE 2.2. Graph showing sulfate deposition, from both poles, corresponding to the 1809 and 1815 eruptions. Because these two major eruptions—Tambora's being "colossal"— occurred within six years, they ensured significant suppression of global temperatures across the entire decade. (Jihong Cole-Dai et al., "Cold Decade [AD 1810–19] Caused by Tambora [1815] and Another [1809] Stratospheric Volcanic Eruption," *Geophysical Research Letters* 36 [2009]: L22703; © American Geophysical Union.)

morning May temperatures slipped 5 degrees below freezing.[13] Such extreme, unseasonal cold was unheard of and would not be felt again until the disastrous volcanic summer of 1816.

A flurry of research since the discovery of the 1809 Unknown has resulted in the identification of the 1810–19 decade as a whole as the coldest in the historical record—a gloomy distinction. A 2008 modeling study concluded Tambora's eruption to have had by far the largest impact on global mean surface air temperatures among volcanic events since 1610, while the 1809 Unknown ranked second over that same period, measuring just over half Tambora's decline.[14] Two papers published the following year confirmed the status of the 1810s as "probably the coldest during the past 500 years or longer," a fact directly attributable to the proximity of the two major tropical eruptions.[15]

In light of this evidence, we can now be sure that the background climate conditions for Tambora were already unusually cool. Its spectacular eruption then increased that cooling to a truly dire extent, contributing to an overall decline of global average temperatures of 1.5°F across the entire decade. One-and-a-half degrees might seem a small number, but as a *sustained* decline characterized by a sharp rise in extreme weather events—floods, droughts, storms, and summer frosts—the chilled global climate system of the 1810s had devastating impacts on human agriculture, food supply, and disease ecologies, as we shall see unfold in often horrific detail in the following chapters.

Just how gloomy were those volcanic years? The Scottish meteorologist George Mackenzie kept meticulous records of cloudy skies between 1803 and 1821 over various parts of the British Isles. Where lovely clear summer days in the earlier period 1803–10 averaged over twenty, in the volcanic decade 1811–20 that figure dropped to barely five. Even that parlous average would have been lower but for the merciful return to seasonal conditions in 1819, which the poet John Keats immortalized for its "mellow fruitfulness" in his poem "To Autumn." For 1816, the Year without a Summer, Mackenzie recorded *no clear days at all*.[16]

A craze for clouds developed during this relentlessly overcast, stormy period. The poetry of the 1810s (notably Shelley's) is full of meditations on "Cloudland, gorgeous land!"—as Coleridge called it.[17] Likewise in painting, by the end of that gray decade, John Constable had given up his impressions of English landscapes for experimental canvases filled entirely with clouds and their subtle peregrinations across a muted sky.[18] The 1810s also marks Charles Dickens's first decade of life. His deep body memory of a volcanic childhood infuses his fiction: think of the "cold, biting weather: foggy withal" of *A Christmas Carol*. Though his adult life and writing career date from a sunnier, warmer England (everything's relative!), his grim weatherscapes of the 1810s have entered the popular imagination as definitive representations of the ever-cloudy, bone-chilling atmosphere of Victorian London.

For the Tamboran decade of the 1810s to warrant the title of "longest sustained cold period" since the Middle Ages is no small thing, since the period 1250–1850 itself has long been referred to as the "Little Ice Age." Before 1250—during the so-called Medieval Warming Period— Englishmen produced their own wine, while the Danes set up farming colonies on Greenland. From the late thirteenth century onward, however, such luxuries were denied by frequent spikes of brutally cold conditions. The English pulled up their vineyards and took to skating on the Thames. No universal glaciation occurred, of course, and bursts of warmth, sometimes decades long, interrupted the general cooling trend. Not an "Ice Age" at all, then—but more like an intermittent six-century cold snap.

Climate models have shown that the cool conditions of the Little Ice Age lie outside the range of natural variability, so climatologists have been compelled to seek out its anomalous causes. One long-popular school of thought has focused on the irregularity of solar radiation, traceable in the historical record through accounts of sunspot activity. Although recent NASA studies support a minor link between solar minima and cooler winters, there have long been skeptics of a solar trigger for the Little Ice Age.[19] The renowned climatologist Alan Robock, among others, has shown that fluctuations in sunspot activity have little impact on global climate on decadal, let alone centennial, scales.[20] According to Robock, as well as a 2012 paleoclimatological study in the Canadian Arctic and Iceland, responsibility for launching the Little Ice Age lies instead with a spasm of major volcanic eruptions

in the late thirteenth century, possibly beginning with a massive tropical eruption in 1258 (precise location again "unknown"). This concentrated sequence of volcanic blasts in the late 1200s altered the baseline
conditions of global climate by a degree or more, beyond the threshold
at which colder temperatures became self-reinforcing through expansion of the Arctic ice pack: a classic climate change feedback loop.

Individually, volcanoes of the magnitude that sustained the Little Ice
Age have the capacity to influence climate for two to three years, until
their aerosol cloud washes from the atmosphere. Volcanoes erupting
in clusters, however—as they did in the thirteenth century and in the
Tambora period of the early nineteenth century—achieve a cumulative
chilling power over global climate by virtue of the slow thermal recovery of the world's oceans, which continue to depress temperatures for
a decade or more after the volcanic dust of any one eruption has vanished from the atmosphere.[21] Thus regular eruptions during the 1250–
1850 period reinforced the initial cooling events. Six centuries after the
1258 Great Unknown, with global volcanic activity on the wane and
temperatures on the upswing, the celebrated Krakatau eruption, which
grabbed world attention in 1883, thus stands as a last hurrah, or encore,
of the Little "Volcanic" Ice Age.

In the chilling case of the volcanic 1810s, the global ocean-atmospheric
system had not yet recovered from the cooling effect of the 1809 Unknown event when the colossal eruption of Tambora occurred. The
aftermath of that eruption, spanning the second half of the decade,
stands as the most catastrophic sustained weather crisis of the millennium. The year following Tambora, 1816, has long enjoyed the folkloric
status of the "Year without a Summer." But this is faint praise (or blame),
indeed. The celebrity of 1816, as a year apart, obscures the greater climatological and social history of the 1810s, of which Tambora's eruption is
the explosive centerpiece. Nor should we view Tambora's global impact
"merely" as an extreme weather regime limited to the late 1810s. Just as
the influence of volcanism on climate may extend centuries, as in the
case of the Little Ice Age, so the social changes wrought by climatic
upheaval on the scale of 1815–18 may be traced decades into the future,
as the following chapters will show.

Fast forward to the twentieth century, and we find only Alaska's Mount Katmai in 1912 and Mount Pinatubo in the Philippines in 1991 erupting on a scale so frequently reached during the Little Ice Age. Even mighty Pinatubo, the century's largest volcanic event, rates an order of magnitude smaller than Tambora, lower even than Tambora's little brother, the 1809 Unknown.[22] Living in New York as I was during the snowy winter of 1991–92, Pinatubo's grip on global climate felt sharp enough! Just arrived from balmy southern Australia, battling the snow and frigid wind in my wholly inadequate outerwear, I wondered how it had occurred to anyone to establish a civilization in such an inhospitable place. In the Tambora year 1816, many residents of New England and the Atlantic seaboard came to the same conclusion, with profound consequences for the history of the United States (a story that must wait for chapter 9).

SCALING TAMBORA

Tambora, in its decapitated state, stakes a serious claim as the most destructive volcano in human history. In light of this, the celebrity of Krakatau's more modest eruption in 1883 seems undeserved. Only the historical accident of the telegraph's invention allowed news of it to travel almost instantly across the globe. Perhaps now, with its bicentenary upon us, Tambora will at last achieve a popular recognition equal to its vaunted reputation among scientists.

But how big was the Tambora event on a geological timescale? According to the Global Volcanism Program at the Smithsonian Institution, the so-called Holocene Period—the approximately twelve thousand years since the last glacial epoch—marks the "official" time frame of volcanic history. Not coincidentally, the consistent, more moderate temperatures characteristic of the Holocene—which have not fluctuated more than a degree in warmth or cold in those twelve millennia—have witnessed great leaps forward for the human race, from a precarious existence in nomadic hunter bands, to literate settler farmer communities, to the advanced technological mass societies of the twenty-first century—a planet literally teeming with humanity.

According to the methodology employed by the Smithsonian—which measures magmatic output deduced from geological sediments and historical accounts—Tambora belongs to an elite group of some half-dozen Holocenic volcanoes that rate 7 on the Volcanic Explosivity Index (VEI), a size officially termed "colossal."[23] Volcanologists are quick to concede the limitations of the index, which gets fuzzier the further back in time one goes. Many important volcanoes are simply missing from the Smithsonian tables, including the great "Unknowns" of 1258 and 1809. Furthermore, paleoclimatologists complain, the VEI measures explosive power only, *not* the climatic impact of the volcano. The famous Mount St. Helens eruption in 1980, for example, rates a respectable 3 on the VEI, but since its volcanic matter ejected sideward and not upward into the stratosphere, its impact on climate was nil. As we have seen, the latitude of volcanoes is also crucial to their climatic impact: tropical volcanoes are capable of influencing global weather patterns while high-latitude volcanoes, such as Iceland's Laki in 1783, impact only the northern hemisphere. But the VEI takes no account of latitude.

Hence the recent development of an alternative index of volcanic measurement based on ice core data—the paleoclimatologist's time travel machine. Historical accounts might reach back centuries, tree-ring data a little further, and geological evidence never see the light of day, but the new Ice-Core Volcano Index (IVI), through identification of sulfur isotope anomalies in polar and mountain ice, offers a trans-millennial measure of ash deposition in the earth's remote glacial archives.[24] Call it the volcanic Hall of Fame. Only eruptions that obtain stratospheric height, and hence climatic significance, find their immortal reward in the ice.

The status of Tambora according to the IVI is more complicated than under the old Smithsonian regime. Tambora's ranking as the largest eruption of the second millennium has been challenged by partisans of other major eruptions of the Little Ice Age, including the 1258

Unknown, Mount Kuwae in 1452, and the 1600 eruption of Huayna-putina in Peru. The debate in the volcano-climate community over the relative magnitude of Holocene and Little Ice Age eruptions will surely continue, with fresh claims and counterclaims year by year. For the purposes of this book, however—standing as it does at the ceremonial eve of Tambora's bicentenary and based on the scientific evidence currently available—I invite the reader to think of Tambora's eruption as a thousand-year volcanic event, and among the very largest since human civilization emerged at the dawn of the Holocene twelve thousand years ago.

After years studying Tambora, when I ponder its relation to other volcanoes, my thoughts do not turn to Vesuvius, destroyer of Pompeii, nor to the great 1258 Unknown, purported trigger of the "Little Ice Age," nor to Krakatau, volcanic darling of the Victorians. I think instead of a volcanic legend with far deeper historical DNA. The cataclysmic eruption of Santorini in the Aegean Sea in 1628 BC has been linked to the collapse of Minoan civilization, the legend of Atlantis, and the Israelites' exodus from plague-ridden Egypt as told in the Bible.[25] An awe-inspiring list, but the tale of Tambora will more than match it. The chapters that follow make the case for Tambora's 1815 eruption as a world-historical event on the order of Santorini's explosion three-and-a-half millennia ago, wreaking, through sinuous courses, comparable changes on global humanity and the totemic mind.

CHAPTER THREE

"THIS END OF THE WORLD WEATHER"

MONSTERS OF GENEVA

On the eve of the infamous lost summer of 1816, eighteen-year-old Mary Godwin took flight with her lover, Percy Shelley, and their baby for Switzerland, escaping the chilly atmosphere of her father's house in London. Mary's young stepsister, Claire Clairmont, accompanied them, eager to reunite with her own poet-lover, Lord Byron, who had left England for Geneva a week earlier. Mary's other sister, the ever dispensable Fanny, was left behind.

The dismal, often terrifying weather of the summer of 1816 is a touchstone of the ensuing correspondence between the sisters. In a letter to Fanny, written on her arrival in Geneva, Mary describes—in hair-raising language that would soon find its way into her novel *Frankenstein*—their ascent of the Alps "amidst a violent storm of wind and rain." The cold was "excessive" and the villagers complained of the lateness of the spring. On their alpine descent days later, a snowstorm ruined their view of Geneva and its famous lake. In her return letter, Fanny expresses sympathy for Mary's bad luck, reporting that it was "dreadfully dreary and rainy" in London too, and very cold.[1]

Stormy northeasters are standard features of Genevan weather in summertime, careening from the mountains to whip the waters of the lake into a sirocco of foam. Beginning in June 1816, these annual storms attained a manic intensity not witnessed before or since. Mary's famous second letter to Fanny is one of the most vivid documents we have of the violent volcanic weather in the Swiss summer of 1816: "An almost perpetual rain confines us principally to the house," she writes on the first of June from Maison Chappuis, their rented house on the shores of Lake Geneva: "One night we *enjoyed* a finer storm than I had ever before beheld. The lake was lit up—the pines on Jura made visible, and all the scene illuminated for an instant, when a pitchy blackness succeeded, and the thunder came in frightful bursts over our heads amid the blackness."[2] A diarist in nearby Montreux compared the bodily impact of these deafening thunderclaps to a heart attack.[3]

The year 1816 remains the coldest, wettest Geneva summer since records began in 1753. That unforgettable year, 130 days of rain between April and September swelled the waters of Lake Geneva, flooding the city. Up in the mountains the snow refused to melt.[4] Clouds hung heavy, while the winds blew bitingly cold. In some parts of the inundated city, transport was only possible by boat. A cold northwest wind from the Jura mountains—called *le joran* by locals—swept relentlessly across the lake. The Montreux diarist called the persistent snows and *le joran* "the twin evil genies of 1816."[5] Tourists complained they couldn't recognize the famously picturesque landscape because of the constant wind and avalanches, which drove snow across vast areas of the plains.

On the night of June 13, 1816, the Shelleys' splendidly domiciled neighbor, Lord Byron, stood out on the balcony of the lakeside Villa Diodati to witness "the mightiest of the storms" that he—well-traveled aristocrat that he was—had ever seen. He memorialized that tumultuous night in his wildly popular poem *Childe Harold's Pilgrimage*:

> The sky is changed—and such a change! Oh night,
> And storm, and darkness, ye are wondrous strong . . .
> From peak to peak, the rattling crags among

FIGURE 3.1 and **3.2**. On the left, a sketch portrait of Mary Shelley at age eighteen, while she was living in Geneva. Several full portraits of a young Mary Shelley were taken in this period, but this tantalizing sketch is the only image to survive (though doubts linger as to its authenticity). It reminds us of how young the author of *Frankenstein* was in the summer of 1816. (Russell-Coates Gallery, Bournemouth; © Bridgeman Art Library.) On the right, Percy Bysshe Shelley as he was in 1819, at age twenty-eight. (By Alfred Clint; © National Portrait Gallery.)

> Leaps the live thunder! Not from one lone cloud,
> But every mountain now hath found a tongue . . .
> How the lit lake shines, a phosphoric sea,
> And the big rain comes dancing to the earth!
> And now again 'tis black,—and now, the glee
> Of the loud hills shakes with its mountain-mirth,
> As if they did rejoice o'er a young earthquake's birth.[6]

In Byron's imagination, the Tamboran storms of 1816 achieve volcanic dimensions—like an "earthquake's birth"—and take delight in their destructive power.

What caused the terrible weather conditions over Britain and western Europe in 1816–18? Why so much rain and so many destructive gales?

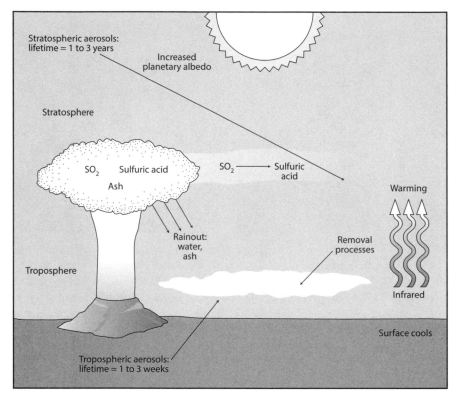

Stratospheric aerosols:
lifetime = 1 to 3 years

Increased
planetary albedo

Stratosphere

SO₂   Sulfuric acid   SO₂ ──→ Sulfuric
Ash   acid

Warming

Rainout:
water,
ash

Removal
processes

Troposphere

Infrared

Surface cools

Tropospheric aerosols:
lifetime = 1 to 3 weeks

FIGURE 3.3. This diagram shows the penetration of loftier volcanic matter into the stratosphere. There, as volcanic sulfur dioxide is chemically transformed into sulfuric acid, an
aerosol layer forms, reducing incoming radiation from the sun and cooling the surface,
even as the stratosphere itself is warmed. (Adapted from M. Patrick McCormick et al.,
"Atmospheric Effects of the Mt. Pinatubo Eruption," *Nature* 373 [February 2, 1995]: 400;
© Macmillan Publishers Ltd.)

The relation between volcanism and climate depends on eruptive
scale. Volcanic ejecta and gases must penetrate skyward high enough to
reach the stratosphere where, in its cold lower reaches, sulfate aerosols
form. These then enter the meridional currents of the global climate
system, disrupting normal patterns of temperature and precipitation
across the hemispheres. Tambora's April 1815 eruption launched enormous volumes of long-suppressed volcanic rock and gases more than
40 kilometers into the stratosphere. This volcanic plume—consisting

of as much as 50 cubic kilometers of total matter— eventually spread across one million square kilometers of the Earth's atmosphere, an aerosol umbrella six times the size of Mount Pinatubo's 1991 cloud.

In the first weeks after Tambora's eruption, a vast volume of coarser ash particles—volcanic "dust"—cascaded back to Earth mixed with rain. But ejecta of smaller size—water vapor, molecules of sulfur and fluorine gases, and fine ash particles—remained suspended in the stratosphere, where a sequence of chemical reactions resulted in the formation of a 100-megaton sulfate aerosol layer. Over the following months, this dynamic, streamer-like cloud of aerosols—much smaller in size than the original volcanic matter—expanded by degrees to form a molecular screen of planetary scale, spread aloft the winds and meridional currents of the world. In the course of an eighteen-month-long journey, it passed across both south and north poles, leaving a telltale sulfate imprint on the ice for paleoclimatologists to discover more than a century and a half later.

Once settled in the dry firmament of the stratosphere, Tambora's global veil circulated above the weather dynamics of the atmosphere, comfortably distanced from the rain clouds that might have dispersed it. From there, its planet-girdling aerosol film continued to scatter short-wave solar radiation back into space until early 1818, while allowing much of the longwave radiant heat from the earth to escape. The resultant three-year cooling regime, unevenly distributed by the currents of the world's major weather systems, barely affected some places on the globe (Russia, for instance, and the trans-Appalachian United States) but precipitated a truly drastic 5–6°F seasonal decline in other regions, including Europe.

The first extreme impact of a major tropical eruption is felt in raw temperature. But in western Europe, biblical-style inundation during the 1816 summer growing season wrought the greatest havoc. To understand the altered precipitation patterns fostered by volcanic weather, we must first grasp the principles of general circulation of the atmosphere. Because of the tilt of the Earth in relation to the sun and the different heat absorption rates of land and sea, solar insolation of the planet is irregular. Uneven heating in turn creates an air pressure gradient across

FIGURE 3.4. *Weymouth Bay*, 1816 (oil on canvas), John Constable (1776–1837). The Victoria & Albert Museum, London, UK. Courtesy of The Bridgeman Art Library.

the latitudes of the globe. Wind is the weatherly expression of these temperature and pressure differentials, transporting heat from the tropics to the poles, moderating temperature extremes, and carrying evaporated water from the oceans over the land to support plant and animal life. The major meridional circulation patterns, measuring thousands of kilometers in breadth, transport energy and moisture horizontally across the globe, creating continental-scale weather patterns. Meanwhile, at smaller scales, the redistribution of heat and moisture through the vertical column of the atmosphere produces localized "weather" phenomena, such as thunderstorms.

In the summer after Tambora's eruption, however, the aerosol loading of the stratosphere heated the upper layer, which bore down upon the atmosphere. The "tropopause" that marks the ceiling of the Earth's atmosphere dropped lower, cooling air temperatures and displacing the jet streams, storm tracks, and meridional circulation patterns from their usual course. By early 1816, Tambora's chilling envelope had created a radiation deficit across the North Atlantic, altering the dynamics of the vital Arctic Oscillation. Slower-churning warm waters north of the Azores pumped overloads of moisture into the atmosphere, saturating the skies while enhancing the temperature gradient that fuels wind dynamics. Meanwhile, air pressure at sea level plummeted across the mid-latitudes of the North Atlantic, dragging cyclonic storm tracks southward. Pioneer British climate historian Hubert Lamb has calculated that the influential Icelandic low-pressure system shifted several degrees latitude to the south during the cold summers of the 1810s compared to twentieth-century norms, settling in the unfamiliar domain of the British Isles, ensuring colder, wetter conditions for all of western Europe.[7]

Across Britain, a radical spike in gale-force westerlies saw platoons of rain-bearing clouds march in from the Atlantic month after month—an airy gray army that brought misery to farmers across Britain and the western continent. In a Constable painting from October 1816, Weymouth Bay—a pretty, sheltered cove on England's south coast where the artist was on honeymoon—sits in fragile sunshine under churning gray-black skies. A couple of optimistic beachgoers—possibly Constable and his new bride—look certain to be drenched. Everywhere, the volcanic winds blew hard. The larger Arctic Oscillation, principal driver of northern European weather patterns, had shifted gears to an anomalous positive phase, as if on steroids.

Both computer models and historical data draw a dramatic picture of Tambora-driven storms hammering Britain and western Europe. A recent computer simulation conducted at the National Center for Atmospheric Research (NCAR) in Boulder showed fierce westerly winds in the North Atlantic in the aftermath of a major tropical eruption, while a parallel study based on multiproxy reconstructions of volcanic impacts on European climate since 1500 concluded that volcanic weather drives the increased "advection of maritime air from the North Atlantic," meaning "stronger westerlies" and "anomalously wet conditions over Northern Europe."[8]

Back at the ground level of observed weather phenomena, an archival study of Scottish weather has found that, in the 1816–18 period, gale-force winds battered Edinburgh at a rate and intensity unmatched in over two hundred years of record keeping.[9] In January 1818, a particularly violent storm destroyed the beloved St. John's Chapel in the heart of the city. The slowing of oceanic currents in response to the overall deficit of solar radiation post-Tambora had left anomalous volumes of heated water churning through the critical area between Iceland and the Azores (engine of the Arctic Oscillation), sapping air pressure, energizing westerly winds, and giving shape to titanic storms.

It was in this literally electric atmosphere that the Shelley party in Geneva, with the celebrity poet Byron attached, conceived the idea of a ghost story contest, to entertain themselves indoors during this cold, wild summer. On the night of June 18, 1816—a signature date in literary history—while another volcanic summer thunderstorm raged around them, Mary and Percy Shelley, Claire Clairmont, Byron, and Byron's doctor-companion John Polidori recited the poet Coleridge's recent volume of gothic verse to each other in the candle-lit darkness at the Villa Diodati. In his 1986 movie about the Shelley Circle that famous summer, the controversial British film director Ken Russell imagines Shelley gulping tincture of opium while Claire Clairmont performs fellatio on Byron, recumbent in a chair. Group sex in the drawing room might be implausible, even for the Shelley Circle, but drug taking is very likely, inspired by Coleridge, the poet-addict supreme. How else to explain Shelley's running screaming from the room at Byron's recitation of the psychosexual thriller "Christabel," tormented by his vision of a bare-chested Mary Shelley with eyes instead of nipples?[10]

From such antics as these, Byron conceived the outline of a modern vampire tale, which the bitter Polidori would later appropriate and publish under Byron's name as a satire on his employer's cruel aristocratic hauteur and sexual voracity. For Mary, the lurid events of this stormy night gave literary body to her own distracted musings on the ghost story competition, instituted two nights earlier. She would write a horror story of her own, about a doomed monster brought unwittingly to life during a storm. As Percy Shelley later wrote, the novel

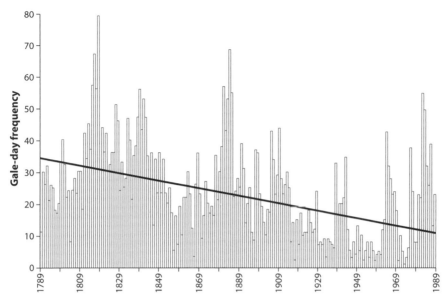

FIGURE 3.5. The number of extreme high-wind days measured in Edinburgh spiked following Tambora's eruption, as this graph clearly shows. The second spike corresponds to the meteorological perturbance following Krakatau's eruption in 1883. (Alastair Dawson et al., "A 200-Year Record of Gale Frequency, Edinburgh, Scotland: Possible Link with High-Magnitude Volcanic Eruptions," *The Holocene* 7.3 [1997]: 339).

itself seemed generated by "the magnificent energy and swiftness of a tempest."[11] Thus it was that the unique creative synergies of this remarkable group of college-age tourists—in the course of a few weeks' biblical weather—gave birth to two singular icons of modern popular culture: Frankenstein's monster and the byronic Dracula.[12]

A week after the memorable night of June 18, Byron and Shelley almost came to grief sailing on Lake Geneva, caught unawares as another violent storm swept in from the east. "The wind gradually increased in violence," Shelley recalled, "until it blew tremendously; and, as it came from the remotest extremity of the lake, produced waves of a frightful height, and covered the whole surface with a chaos of foam." By some miracle they found a sheltered port, where even the storm-hardened locals exchanged "looks of wonder." Onshore, trees had blown down or been shattered by lightning.[13]

The pyrotechnical lightning displays of June 1816 ignited the literary imaginations of Mary Shelley and Lord Byron. In perhaps the most famous stanza of *Childe Harold's Pilgrimage*—"Could I embody and unbosom now / that which is most within me"—Byron defines emotional "expression" itself by the single word, "*Lightning.*" Likewise in *Frankenstein*, Mary Shelley uses the experience of a violent thunderstorm as the scene of fateful inspiration for her young, doomed scientist:

> When I was about fifteen years old . . . we witnessed a most violent and terrible thunder-storm. It advanced from behind the mountains of Jura; and the thunder burst at once with frightful loudness from various quarters of the heavens. I remained, while the storm lasted, watching its progress with curiosity and delight. As I stood at the door, on a sudden I beheld a stream of fire issue from an old and beautiful oak, which stood about twenty yards from our house; and so soon as the dazzling light vanished, the oak had disappeared, and nothing remained but a blasted stump.[14]

Frankenstein's life is changed in this moment; thenceforth he devotes himself, with maniacal energy, to the study of electricity and galvanism. In the fierce smithy of that Tamboran storm, Frankenstein is born as the anti-superhero of modernity—the "Modern Prometheus"—stealer of the gods' fire.

THE FIRST METEOROLOGIST

Jeff Masters, a professor of meteorology at the University of Michigan, is perhaps the most-read weather blogger in the United States. In an extensive posting in June 2011, he reflected on the wave of apocalyptic storms, floods, and droughts of the previous twelve months across the United States and the globe as probably the most tumultuous planetary sequence of extreme weather events since 1816.[15] That Masters, a preeminent meteorologist and historian of our own era of climate deterioration, would consider 1816 the baseline example for global "extreme weather" in the twenty-first century impresses on us the historical scale of the storms that inspired Mary Shelley and her talented circle that legendary summer on the shores of Lake Geneva. Indeed, in the dozen millennia since the retreat of the glaciers opened the door to human civilization, people have rarely, if ever, seen weather like it.

The folkloric history of 1816's extreme weather, especially in Europe and North America, looms large in the minds of meteorologists. That the myriad legends of the "Year without a Summer" have some statistical basis in the temperature record is owed—at least in England—to a man with a strong claim to the title of "father of meteorology": the austere Quaker from Tottenham, Luke Howard. Howard's landmark publication, *Essay on the Modification of Clouds* (1803), introduced the basic cloud classifications—cirrus, nimbus, and so forth—still in use today. The essay inspired fan mail from the German poet-scientist Goethe, while in 1813 Thomas Forster credited Howard with "the *daily* increasing attention devoted to this science."[16]

The first wave of modern European interest in meteorology, with Luke Howard at the vanguard, coincided with the historically cold and cloudy decade of the 1810s.[17] From 1807 to 1819, Howard maintained the first professional almanac of British weather conditions, complete with detailed statistical tables and prolific commentary. Its very title, *The Climate of London*, proclaimed the first principle of modern climatology: that "climate" is the aggregate of weather conditions in a particular region *over time*, to be distinguished—as a legitimate science—from the vulgar gossip, anecdote, and superstition that traditionally surround the weather (a discourse as "dreary" as the weather itself is so often complained to be).

Howard's *Climate of London* offers hard evidence of the altered weather patterns across western Europe produced by Tambora's eruption. By the first week of January 1816, Howard was recording "gales" and "violent storms of wind and rain" in London and elsewhere on a near daily basis. Provincial correspondence brought accounts of never-before-seen storm activity. The winter also brought the first indications of the historical cold temperatures that would afflict the country through 1816 and beyond. "I had . . . opportunity of observing at Totten-ham," Howard wrote,

> the *intense* cold of the 9–10th of the second month, 1816 . . . a gale from
> the North East had precipitated in snow the moisture which previously
> abounded. . . . So cold was the surface on the 9th at noon, that a bright sun,
> contrary to its usual effect in our climate, produced not the least moisture
> in the snow, the polished plates of which retaining their form, refracted the
> rays with all the brilliancy of dew drops.[18]

The daytime temperature on February 9, 1816, never exceeded 20°F, slip-ping that night to 5 below zero and remaining there for twelve hours. It was a phenomenon of cold "not uncommon" in higher latitudes, How-ard commented, but truly remarkable for the south of England. The sun shone but had seemingly lost its power of warmth.

The subsequent summer of 1816—that would live in infamy—brought a continuation of storms, gales, and cold conditions. Amazed locals reported snows on the summit of Helvellyn in northern England in July and snowdrifts five feet deep in the north of Scotland. Picking up the newspaper, Howard read naval reports in July of conditions at sea seeming more like the worst of a wild winter, including "strong gales, ships on shore, and [the] loss of anchors." For Britons, the summer of 1816 was shaping up as a full-blown weather emergency: "From all parts of the country we hear of damage done by the late storms, and floods occasioned by the heavy rains."[19]

Like many Englishmen of means during the summer of 1816, Luke Howard took advantage of the long-awaited end of the Napoleonic Wars to travel through Europe, off-limits to tourists for two decades. It was a busman's holiday for Howard, whose meteorologist's eye was awestruck by the continental scale of the 1816 climate crisis:

> From Amsterdam to Geneva, I had ample occasion to witness the fact that
> the excessive rains of this summer were not confined to our own islands, but
> took place over a great part of the continent of Europe. From the sources
> of the Rhine among the Alps, to its embouchure in the German ocean, and

through a space twice or thrice as broad from east to west, the whole season presented a series of storms and inundations.

Everywhere he went, Howard saw villages under water and entire neighborhoods of large cities flooded. He came upon dikes destroyed and bridges reduced to ruins by flash floods. He rode by vast fields of submerged crops and others simply borne away by torrents of water that flowed relentlessly in all directions, transforming the pleasant tourist geography of agrarian Europe in summertime into a continent-wide disaster zone.

Given the biblical flooding before his eyes, Howard was amazed to learn that to the north, in Scandinavia, farm fields lay "parched with drought" and that churches in Danzig and Riga were holding night-long prayer vigils for rain. By shifting the latitudinal patterns of precipitation and intensifying weather systems across the board, Tambora brought both flood and drought to the Europe of 1816–18, a pattern we will see repeated around the globe.

Passing through Switzerland, Howard traveled the same scenic routes taken by Mary Shelley and her circle of friends. While Byron and the Shelleys exchanged ghost stories, Howard's professional eye was drawn to the startling summer accumulation of snow on even the lower elevations of the alpine mountains:

> I saw the snows of the preceding winter lying in very large masses, in hollows on the chain of the Jura, and on the Mole near Geneva, from whence they usually vanish in summer; and this at a time when the new snows had already begun to fall on the same summits.

Back in England in the autumn of 1816, Howard recorded more apocalyptic weather. Around lunchtime on October 7, he experienced a "loud explosion of electricity"—a bolt of lightning—that shook the ground at Tottenham for several seconds. "Thunder in long peals and vivid lightning" then continued for more than an hour. On November 6, a dense cloud of Tambora's volcanic dust enveloped Chester in the west of England. At noon, amid impenetrable darkness, citizens of the cathedral town lit candles and carried lanterns through the streets. Hail, frost, and snow two feet deep followed in the succeeding days. The same conditions prevailed over London later in the month, where Howard recorded a noontime temperature of 2°F and the daytime darkness required coachmen to dismount to light the way for their horses.

The creeping terror inspired by Tambora's unnatural weather regime was due to its unrelenting delivery of extreme conditions. Entering the second winter after Tambora's explosion, Howard continued to gather reports of storm systems of "a severity almost beyond example." In December, he listed hailstorms, gales of "an excessive degree of violence," and earth tremors caused by lightning—just in Tottenham. Like the painter Turner, he also noted what was, unbeknown to him, the startling effect of Tambora's aerosol cloud on the atmospheric spectrum. On December 27th, in the midst of storm clouds, the setting sun appeared before him like an angry giant, "fiery red, and much enlarged."

With the cold, wild year of 1816 at last at an end, Howard was able to assess its severity on a hard statistical basis. The results must have shocked even this mild-mannered Quaker and put him in mind of the vengeance of the Lord. In his previous nine years of temperature observations, 1807–15—an already below-average sample owing to the impact of the 1809 Unknown eruption—the average daily temperature in London had been 50°F. In 1816, the average fell by 12 degrees, to 38°F.[20] The "Year without a Summer" appears too mild a description for the meteorological annus horribilis that was 1816. More like the "Year without a Sun."

In the pre-Tambora sections of his *Climate of London*, Howard's interests are distinctly parochial, limited to weather observations in the British Isles and greater London in particular. Following his firsthand experience of volcanic weather conditions in continental Europe, however, Howard takes care to keep track of reports from abroad. His 1817 almanac lists "hurricanes" in Hamburg and Amsterdam, hailstorms across France, "excessive cold" in Lisbon, and continued "inundations" in Switzerland. In the widened horizons of an amateur weather enthusiast in 1816–17, then, we see the origins of modern synoptic meteorology, which understands weather as a cross-continental phenomenon and not simply the variation of local conditions.

In Germany, another budding meteorologist—the polymath Heinrich Brandes—had arrived at the same conclusion about the broader geographical scales of weather. Out of the trauma of middle Europe's "Year of the Beggar"—and no doubt humiliated at the destruction by flood of dykes he himself designed on the River Weser in Lower Saxony—Brandes promoted a continental overview of weather patterns. Looking back over the disastrous year, Brandes argued in a letter dated December 1816 that "more precise reports of the weather, even if only for the whole of Europe, would surely yield very instructive results. If one could draw maps of Europe according to the weather for all 365 days of the years, then it would of course show, for instance, where the boundary of the great rain-bearing clouds, which in July covered the whole of Germany and France, lay."[21] Acting on this notion, Brandes began to design and compile the world's first weather maps, published in 1820, the same year as Howard's landmark *Climate of London*." Call it intellectual teleconnection. Tambora's eruption was a mother to global suffering on an epochal scale; but it must count, likewise, as a major birth event in modern meteorology. Catastrophic climate change generates world-changing ideas as well as global-scale trauma.

Howard's observations of wild weather continue through 1817. Hailstones "large as hazelnuts" and "pigeon's eggs" rained down through the summer while, for the third straight year, winter storms descended upon the British Isles with millennial ferocity, concluding with an epic tempest on March 4, 1818, that cut a violent swath across the south of England. Included in the destruction was a famous tree in Plymouth, the newspaper account of which reads very much like the lightning strike passage in *Frankenstein*, first published the month of this memorable storm:

> The effects of the late thunderstorm [were] the most extraordinary that ever occurred in this county.... The tree in question has long been admired for its girth and noble proportions, being more than 100 feet high and nearly 14 feet in girth; but it exists no longer, having been literally shivered to

pieces by the electric fluid. Some of the fragments lie 260 feet from the spot, and others bestrew the ground in every direction, presenting altogether a scene of desolated vegetation, easier to be conceived than described.[23]

The following month—Tambora's three-year anniversary—another bizarre spring storm raged across the environs of London, unroofing houses and blowing down walls. On Hampstead Heath—sometime residence of the Shelleys, Coleridge, Keats, and painter Constable—dozens of trees were blown up from their roots.

The litany of climate change destruction post-Tambora is exhausting to read and almost numbingly repetitive. So many trees blown down, fields flooded, and crop-killing frosts and snows. But unlike twenty-first-century climate change, which has no end in sight but only a relentless acceleration, Tambora's weather emergency did eventually pass. By June 1818, Luke Howard was able to report "clear hot sunshine" that brought with it the warmest stretch of weather since 1808, and "a period unequalled in dryness since the beginning of 1810." Walking the semi-rural environs of Tottenham, Howard observed "the deeper green of the foliage and the richer colour of many flowers."[24] With Tambora's global dust cloud lifting, it seemed like the refreshment of the world. The *greenness* of leaves and flowers at full blush—everyday glories almost forgotten—now resumed for Howard their wonted niche in the cherished visible world. Then in the autumn, the tender nasturtiums and horse chestnuts returned like old friends, untormented by early frosts. What a soul-lifting sight it must have been for the observant Quaker, attentive to signs of God's mercy.

EUROPE'S LAST FAMINE

Tambora's shaping influence on human history does not derive from extreme weather events considered in isolation but in the myriad environmental impacts of a climate system gone haywire. As I have argued, the popular moniker awarded 1816, the "Year without a Summer," sounds altogether too benign, no more than the inconvenience of donning an overcoat in July, when, in fact, "no summer" meant "no food" for millions of people. As a result of the prolonged poor weather, crop yields across the British Isles and western Europe plummeted by 75% and more in 1816–17. Tambora's calling card in Germany, where it is remembered as the "Year of the Beggar"—or in Switzerland, as "L'Année de la misère" and "Das Hungerjahr"—better captures the atmosphere of social crisis during the extreme weather onslaught of 1815–18. In the first summer of Tambora's cold, wet, and windy regime—the "atmospheric sarabande" of 1816—the European harvest languished miserably. Farmers left their crops in the field as long as they dared, hoping some fraction might mature in late-coming sunshine. But the longed-for warm spell never arrived and at last, in October, they surrendered. Potato crops were left to rot, while entire fields of barley and oats lay blanketed in snow until the following spring.

In Germany, the descent from bad weather to crop failure to mass starvation conditions took a frighteningly rapid course. Carl von Clausewitz, the military tactician, witnessed "heartrending" scenes on his horseback travels through the Rhine country in the spring of 1817: "I saw decimated people, barely human, prowling the fields for half-rotten potatoes."[25] In the winter of 1817, in Augsburg, Memmingen, and other German towns, riots erupted over the rumored export of corn to starving Switzerland, while the locals were reduced to eating horse and dog flesh.[26]

Meanwhile, back in England, riots broke out in the East Anglian counties as early as May 1816. Armed laborers bearing flags with the slogan "Bread or Blood" marched on the cathedral town of Ely, held its magistrates hostage, and fought a pitched battle against the militia.[27] In Somersetshire, three thousand coal miners took over the local mine in their desperation over sky-high bread prices. When asked what they wanted, they replied, "full wages, and that they were starving." The local magistrate responded by reading the Riot Act, threatening all malingerers with death, and sending in the militia to attack the crowd with "immense bludgeons."[28] On an even larger scale, in March 1817 more than ten thousand demonstrated in Manchester while in June the so-called Pentrich Revolution involved plans to invade and occupy the city of Nottingham. The army was called in to quell similar disturbances in Scotland and Wales. The government of Lord Liverpool responded to the desperation of the people with draconian force. It suppressed publication of agricultural quarterly reports and suspended habeas corpus. Provincial jails filled to overflowing across the kingdom, while scores of hungry rioters were hanged or transported overseas to penal colonies.[29]

In his magisterial account of the social and economic upheaval in Europe during the Tambora period, historian John Post has shown the scale of human suffering to be worst in Switzerland, home to Mary Shelley and her circle in 1816. Even in normal times, a Swiss family devoted at least half its income to buying bread. Already by August 1816, bread was scarce, and in December, bakers in Montreux threatened to cease production unless they could be allowed to raise prices.[30] Subsequently, when the price of grain almost tripled in 1817, a basic subsistence diet was suddenly out of the reach of hundreds of thousands of Swiss, not only in the agricultural regions but in industrial towns where "the weekly earnings of a hand-spinner in 1817 . . . were less than the price of a pound of bread."[31] With imminent famine came the threat of "soulèvements": violent uprisings. Bakers were set upon by starving mobs in the market towns and their shops destroyed. The English ambassador to Switzerland, Stratford Canning, wrote to his prime minister that an army of peasants, unemployed and starving, was assembling to march on Lausanne.

As historical coincidence would have it, Stamford Raffles spent that woeful summer traveling in Europe, having left his governor's post in Java. He thus holds the dubious distinction of being the only person to leave an account of both Tambora's eruption in the Dutch East Indies in 1815 and the subsequent years of extreme weather and famine in Europe, on the other side of the world. To Raffles and his brother Thomas, who traveled with him and kept a diary, the provincial villages of France appeared like ghost towns:

> We could not but notice the almost total absence of life and activity. . . .
> There was an air of gloom and desertion pervading them. The houses had

a cheerless and neglected appearance. No one was seen in the streets—they looked as if deserted by their population.

Across France, grain crops rotted in the rain-soaked fields, while vintners in 1816 gathered the most meager grape harvest in centuries of record keeping. Crossing into landlocked Switzerland, where grain prices rose two to three times higher than in the coastal regions, the Raffles brothers found the food shortage even more dire: "the great increase of beggars . . . chiefly children . . . was truly astonishing."[32]

In dealing with the crisis, the Swiss authorities were disadvantaged by a fragmented political structure. When serious dearth threatened, administrators of the tiny cantons panicked and closed their frontiers to the export of grain, ensuring they themselves would be unable to import emergency supplies. Public works programs and soup kitchens averted a greater calamity, but thousands still died of starvation during continental Europe's "last great subsistence crisis." A priest from Glaris painted a lamentable portrait of the suffering in his district: "It is terrifying to see these walking skeletons devour the most repulsive foods with such avidity: the corpses of livestock, stinking nettles—and to watch them fight with animals over scraps."[33] Everywhere, desperate villagers resorted to a pitiful famine diet of "the most loathsome and unnatural foods."[34] Mortality in 1817 was over 50% higher than its already elevated rate in the war year 1815, while deaths exceeded births in Switzerland in both 1817 and 1818, suggesting an excess mortality rate in the tens of thousands. Only intermittent, timely shipments of grain from Russia, which had fortuitously escaped the worst post-Tambora weather conditions, prevented Switzerland and much of Europe from collapsing into full-scale famine.

Conditions were desperate enough even so. To tourists on the European continent in 1817, the legions of vagrant poor descending upon the market towns appeared like advancing columns of an army on the march. The prefect of Brie described the flood of refugees into his province as like "an invasion or perhaps the migration of an entire nation."[35] In the extremity of their suffering, beggars lost all fear of the law. Waves

of arson, assault, and robbery swept the countryside. The Montreux di-
arist records widespread fears of a return to "scenes from the age of the
Goths and the Vandals."[36] Inevitably, some Swiss authorities overreacted.
Thieves were beheaded and minor pilfery punished with whipping.

Most shocking of all was the fate of some desperate mothers. In horrific circumstances repeated around the world in the Tambora period,
some Swiss families abandoned their offspring in the crisis, while others
chose killing their children as the more humane course. For this crime,
some starving women were apprehended and decapitated. Thousands
of Swiss with more means and resilience emigrated east to prosperous
Russia, while others set off along the Rhine to Holland and sailed from
there to North America, which in 1817–19 witnessed its first significant
wave of refugee European migration in the nineteenth century. The
numbers of European immigrants arriving at U.S. ports in 1817 more
than doubled the number of any previous year.[37]

In political terms, the food shortages and social instability of the
Tambora period spurred governments to the authoritarian, rightward
shift we associate with the ideological landscape of post-Napoleonic
Europe. In the words of Swiss liberal journalist Eusèbe-Henri Gaullieur
(an impressionable boy at the time of the crisis), "The gains made by the
spirit of progressive liberalism were substantially eroded . . . by the suffering arising from the disaster of 1816." Fear of agricultural shortfall also
motivated political leaders to adopt protectionist policies. It is during
the Tambora emergency that tariffs and trade walls first emerged as
standard features of the European and transatlantic economic system.[38]

FRANKENSTEIN AND THE REFUGEES

But it would not do to dwell on the macro-implications of post-Tambora
chaos without giving proper memorial to its principal victims: the
common people who faced the slow torture of death by starvation.
Devastated by famine and disease in the Tambora period, the poor of
Europe hurriedly buried their dead before resuming the bitter fight for
their own survival. In the worst cases, children were abandoned by their

families and died alone in the fields or by the roadside. The well-born members of the Shelley Circle, of course, were never reduced to such abysmal circumstances. With credit to spare, they did not experience the food crises that afflicted millions among the rural populations of western Europe in the Tambora period. That said, chroniclers of Mary Shelley and her friends have been wrong to imagine their European existence as a charmed bubble of poetry, romantic villas, and sexual intrigue. The Shelleys' celebrated writings were very much enmeshed within the web of ecological breakdown following the 1815 Tambora eruption, when a subsistence emergency weakened the European population and famine-friendly contagion took hundreds of thousands of lives.

Byron and Percy Shelley were companions on a weeklong walking tour of Alpine Switzerland in June 1816, during which they debated poetry, metaphysics, and the future of mankind but also found time to remark on the village children they encountered, who "appeared in an extraordinary way deformed and diseased. Most of them were crooked, and with enlarged throats."[39] In Mary's novel *Frankenstein*, conceived that summer, the Doctor's benighted creation assumes a similar grotesque shape: a barely human creature, deformed, crooked, and enlarged. As remarkable a feat of literary imagination as *Frankenstein* is, Mary Shelley was not wanting for real-world inspiration for her horror story, namely, the deteriorating rural populations of Europe post-Tambora.

On the thunderous night in Geneva that first inspired her ghost story, Mary Shelley pictured Frankenstein waking from a nightmare to find his hideous creation at his bedside, "looking on him with yellow, watery, but speculative eyes."[40] The description is reminiscent of Percy Shelley's encounter with "half-deformed or idiotic" beggar children, presumably deranged with hunger. Numerous similar impressions could be cited. Another English tourist, traveling from Rome to Naples in 1817, remarked on "the livid aspect of the miserable inhabitants of this region." When asked how they lived, these "animated spectres" replied simply, "We die."[41] Rural Europe in 1816 descended into a land of the living dead. If their imaginations had not been exhausted by the creation of Frankenstein and the vampire, someone in the Shelley Circle would surely have invented the zombie.

Mary Shelley's imaginative conjuring of her famous Creature thus bears the mark of the famished and diseased European population by which she was surrounded that dire Tambora summer. Like the hordes of refugees spreading typhus across Ireland and Italy during Shelley's writing of the novel, the Creature is a wanderer and a menace to civilized society. At his merest touch, healthy people drop dead like flies. In the novel, this murderous capability is attributed to the monster's preternatural strength. But the terrifying atmosphere of his rampage, and his ability to strike at will across thousands of miles, seems more like the spread of a famine or contagion.

Like the hordes of refugees on the roads of Europe seeking aid in 1816–18, the Creature, when he ventures into the towns, is met with fear and hostility, while the privileged families of the novel, the De Lacys and the Frankensteins, look upon him with horror and abomination. If we look beyond the much-discussed scientific resonances of the monster's creation, the lived experience of Mary Shelley's creature most closely embodies the degradation of the homeless European poor during the Tambora period. The violent disgust of Frankenstein and everyone else toward him likewise mirrors the utter want of sympathy shown by many affluent Europeans toward the millions of Tambora's climate victims suffering hunger, disease, and the loss of their homes and livelihoods. As the indigent Creature himself puts it, he suffered first "from the inclemency of the season" but "still more from the barbarity of man."[42]

"THE BRIGHT SUN WAS EXTINGUISH'D"

In a letter written in the last days of July 1816, Lord Byron complained, as Mary's friends all did, of "stupid mists—fogs—rains—and perpetual density."[43] In that litany of poor weather, however, one depressing day stood out, "a celebrated dark day, on which the fowls went to roost at noon, and the candles lighted as at midnight."[44] This must have been the same sun-canceling cloud reported over Leige on July 5, 1816, as "an enormous mass in the form of a mountain."[45] Most likely it belonged to the Tamboran weather pattern that, reaching into the stratosphere for a chunk of volcanic dust, dumped red snow on the southern Italian town of Taranto in early May, terrifying the inhabitants.[46] Tambora's global dust veil, or a very dense portion of it, had settled directly over western Europe.

From his balcony at the Villa Diodati on Lake Geneva, Byron enjoyed a front-row view of the day Tambora's ash cloud blocked the Alpine sun. As a memorial of that weird event, he wrote a long apocalyptic poem he called "Darkness." Though written from the shell of aristocratic entitlement, Byron's rich, humanistic imagination allowed him to combine the literal atmosphere of doom of that July day in 1816 with speculation on a social landscape transformed by environmental collapse. "Darkness" accordingly stands as a classic meditation on the human impacts of climate change. It begins portentously—

I had a dream, which was not all a dream.
The bright sun was extinguish'd, and the stars
Did wander darkling in the eternal space,
Rayless, and pathless, and the icy earth
Swung blind and blackening in the moonless air;
Morn came, and went—and came, and brought no day,
And men forgot their passions in the dread
Of this their desolation; and all hearts
Were chill'd into a selfish prayer for light (ll. 1–9)

Byron imagines the erosion of human sociability in a toxic, degraded landscape. Traumatized victims of ecological catastrophe suffer a slew of social-emotional disorders, experienced in overwhelming feelings of "dread" and "desolation," injustice and resentment, and a violent "selfishness."

In Byron's "seven sorrows" poem from Tambora's aftermath, we see a thematic trajectory that parallels *Frankenstein*: in the midst of meteorological tumult, human sympathy fails. The "selfish prayer[s]" of the

FIGURE 3.6. Lord Byron on his balcony at the Villa Diodati by Lake Geneva. Here, at perhaps the most famous rented address in British literary history, he played host to the Shelleys in the summer of 1816, witnessed the great storm of June 13, and wrote the apocalyptic poem "Darkness." (Print Collection, Miriam and Ira D. Wallach Division, New York Public Library.)

people lead to social breakdown, violence, and chaos. In the volcanic cooling of 1816, human hearts are "chill'd" along with the atmosphere. This is Byron's *Apocalypse Now*. The fragile edifice of civilization has crumbled—no cities, no agriculture—leaving a traumatized human remnant to wander across a scene of biblical desolation:

> The brows of men by the despairing light
> Wore an unearthly aspect, as by fits
> The flashes fell upon them; some lay down
> And hid their eyes and wept; and some did rest
> Their chins upon their clenched hands, and smiled;
> And others hurried to and fro, and fed

Their funeral piles with fuel, and looked up
With mad disquietude on the dull sky,
The pall of a past world; and then again
With curses cast them down upon the dust,
And gnash'd their teeth and howl'd ... (ll. 22–32)

Birds fall from the sky, animals are massacred, and wars break out—"no
love was left." Then, inevitably, arises the specter of universal Hunger:

All earth was but one thought—and that was death,
Immediate and inglorious; and the pang
Of famine fed upon all entrails—men
Died, and their bones were tombless as their flesh (ll. 42–45)

With a remarkable, prescient sympathy, Byron's "Darkness" antici-
pates the full-blown humanitarian disaster as it was to unfold in Switzerland and around the world over the subsequent three year global
climate emergency. At a time when the modern media and information
age were in their earliest infancy, Byron's apocalyptic fantasy and Mary
Shelley's legendary horror story are notable examples of the European
literate class's symbolic response to the colossal social trauma unfolding around them in the Tambora crisis years of 1816–18. In their unforgettable works, Byron and Shelley imagined the experience of the starving and diseased millions who never enjoyed proper representation in
the press and parliaments of Europe, but mostly sank into oblivion,
unmourned.

THE BOLOGNA PROPHECY

Lord Byron was not alone in his apocalyptic speculations in 1816. In
fact, a fever of the end times swept Europe as virulently as typhus in
the post-Tambora period. In Bologna, Italy, an astronomer predicted the
world would end on precisely July 18, 1816, with the breakup of the
sun. The so-called Bologna Prophecy, reported in newspapers across

Britain and the Continent, became a lightning rod for millennialist panic surrounding the deteriorating weather and political instability in the aftermath of Waterloo.[47] According to the "mad Italian prophet," renewed sunspot activity pointed to an imminent "solar catastrophe" certain "to put an end to the world by conflagration."[48] For others, the sun's unprecedented dimness gave rise to fears the life-giving orb might become "wholly incrusted, so as to plunge us at once into the unutterable darkness that characterized the primitive chaos."[49]

The-End-Is-Nigh cranks are always with us. But in the atmosphere of heightened public concern about the aberrant weather, Bolognese officials took the precaution of throwing the doomsaying astronomer in prison. To no avail. For fifteen days, churches in Belgium were filled with penitents engaged in silent, preparatory prayers. On July 12—another stormy, thundery day—three-quarters of the inhabitants of Ghent (so it was said), on mistaking the martial music of a passing regiment for the trumpet call of the Day of Judgment, ran out onto the streets in loud "lamentation" and "threw themselves on their knees." Meanwhile on the streets of Paris on July 17, one could purchase a pamphlet promising, in large type, "Détails sur la fin du monde!"[50] In the words of one Swiss observer, the fear that a piece of the sun would break off and crash into the earth "gripped all of Europe."[51] Newspaper editors in England zigzagged between gleeful mockery of the whole affair and warning each other not to feed public hysteria.[52] July 18 came and went, but the religious atmosphere of the season did not abate. A week later, the Swedish queen herself presided over a march of six thousand peasants to the cathedral at Bex to pray for deliverance from God's wrath.[53] Back in Highgate, outside London, the poet Coleridge summed up the popular mood in calling the wave of violent summer storms "this end of the world weather."[54]

Such was the chaotic scene across Europe in the first summer after Tambora's eruption. Nor was the atmosphere of hysteria ill-founded. The sun may not have extinguished itself, but tens of thousands of Europeans would lose their lives in the following two years from malnourishment, outright starvation, or famine-driven diseases like typhus.

Hundreds of thousands more were displaced from their homes and communities, left to wander the highways of Europe.

The full dimensions of Tambora's impact on global human society in the nineteenth century, as I have argued from the outset, have never been properly understood. What historical research has been done—by both scholars and popular historians—has been confined to the North Atlantic zone—to Tambora's impact on western Europe and the United States. Elements of the story I have told in this chapter have thus been presented elsewhere, though not synthesized, as I have done here, into a telling case study of the direful social effects of abrupt climate change. I will return to Europe in later chapters with a more detailed narrative of Tambora's aftermath in Ireland and the Alps, before concluding with a history of the Tambora disaster as it unfolded in parts of the United States where this book was in fact written: the Midwest and Mid-Atlantic.

But first we must travel to distant regions where no history of Tambora's deadly grasp has ever been traced. The science on Tambora has been global in its descriptions for decades, while social history, confined to the Euro-American zone, has remained stubbornly parochial until now. From epidemic disease in the Bay of Bengal, to the withered rice fields of Yunnan, to a disintegrating Arctic ice pack, the fingerprints of Tambora's killer plume are to be found worldwide through the late 1810s and beyond. Armed with twenty-first-century scientific instruments—and a global, teleconnected imagination—we are set to break this case wide open.

CHAPTER FOUR

BLUE DEATH IN BENGAL

APOLLO'S DEADLY ARROWS

Homer's epic war poem, the *Iliad*, long honored as the originating text of Western literature, opens with an invading army encampment devastated by disease. The god Apollo, angered by the Greeks' poor treatment of one of his priests, descends on the beach of Troy "angered in his heart":

> He came as night comes down and knelt then
> apart and opposite the ships and let go an arrow.
> Terrible was the clash that rose from the bow of silver . . .
> The corpse fires burned everywhere and did not stop burning.[1]

Given the classical education of the British empire's medical men, it is not surprising that the epic-sized nineteenth-century literature on cholera opens similarly, with the story of Indian governor Lord Hastings's "Grand Army," in late 1817, brought to its knees by a deadly epidemic that seemed to come from the malignant skies. Again and again, British writers on cholera over the coming decades would return to the near-rout of British forces in that Tambora year to comment upon and analyze the cholera, as if it were a famous scene in an Homeric poem or Shakespearean tragedy.

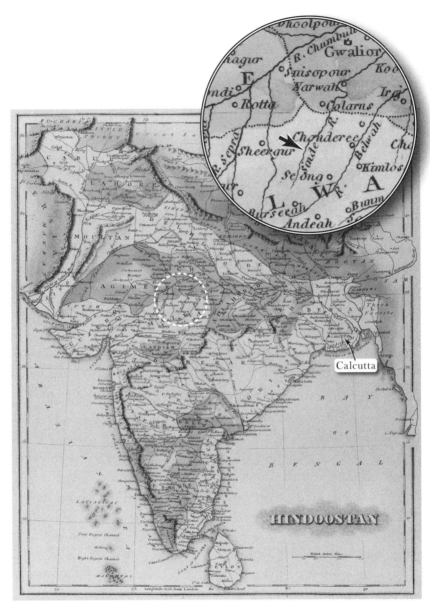

FIGURE 4.1. The Maratha War in 1817 was waged in the independent regions of north-central India, south of Delhi. Lord Hastings's forces, encamped south of the city of Gwalior on the River Sinde, removed east to the River Betwah after the deadly cholera outbreak. (Adapted from a contemporary map by American cartographer Fielding Lucas, in his *General Atlas, Containing Distinct Maps of All the Known Countries in the World* [Baltimore, 1823].)

After a long, hot march out of Bengal, the British Grand Army was encamped by the River Sinde in north-central India. From his position at the base of a thickly wooded range of hills, Hastings could guard the sole eastern access to the fortress town of Gwalior, bastion of the independent Maratha prince Scindia. Cut off from his allies, Scindia had just concluded a tense treaty with Hastings to withhold his support for the roving Pindaree militias, whom the British governor hoped to destroy.

In the early afternoon of November 8, soldiers brought two Indian stretcher-bearers to the tent of Frederick Corbyn, a medical officer attached to the Bengal regiment of Hastings's army. The men's skin was clammy and leaden, their eyes sunken, pulse almost imperceptible. When one of them vomited, the sight of the rice-colored odorless fluid prompted Corbyn to seek out the superintending surgeon, who told him to break camp immediately and find safer ground. Before he could act upon this order, however, an officer of the rear guard arrived to report that hundreds of camp followers and soldiers were already dead or dying along the line of march, their lips and fingers a telltale blue. Accompanied by a cavalry guard, Corbyn hurried to the road, where he found the Indian regiments and their followers in a terrible state. Whole families who, that same morning, he had seen set off in perfect health were lying dead by the streams near the road.[2]

The next day, cholera swept through the camp of the British Grand Army with "indescribable violence." Between November 15 and 20 alone, five thousand men, women, and children died. All military maneuvers ceased, as the camp transformed into a hospital and open-air morgue. An eerie quiet descended, broken only by the groans of the dying. The British kept to their tents, venturing out only to inquire about the state of sick friends, while the Indians bore the biers of their dead to the river in silence. At the height of the epidemic, even these rituals ceased. The victims were thrown into ravines or brought to the English tents and left there—the guilt for their deaths laid ceremonially at the door of the colonizing power. Many Indians blamed the epidemic on the slaughter of a cow to feed the British officers in a nearby grove sacred to Hardaul Lala, the deified ancestor of a local noble family. Hardaul Lala subsequently became one of the new popular cholera deities, with temples as far away as Lahore.[3]

By the time it was reconstituted, Hastings's camp had lost half its numbers. Thirty thousand followers deserted, while some ten thousand died, many of them in the act of deserting. The roads for miles around were littered with the bodies of those who had not been able to outrun the blue death. In the commander's tent, two Indian servants collapsed behind Hastings's chair as he worked, while the general himself gave orders that in the event of his own perishing from the contagion he should be quietly buried inside the tent so as not to demoralize his troops or send hope to Britain's prevaricating Indian allies that the terms of the recent treaty might be disregarded.

Convinced the cause of the epidemic lay in their unhealthy situation by the River Sinde, surrounded by forest and swamp, Hastings ordered the removal of the army to higher ground. This process was hampered by the paralyzed state of the camp and the lack of vehicles with which to transport the sick. Carts, cattle, and elephants were recruited from the nearby villages, but even these proved insufficient. Many were left to die in the road. Hundreds more dropped during each day's march, creating "the appearance of a field of battle and . . . an army retreating under every circumstance of discomfiture and distress." Hastings, confiding his dismay to his journal, called the experience "heartbreaking" and "a most afflicting calamity."[4]

Finally, after a week of halting, harrowing progress, Hastings's division encamped on the heights at Erich by the holy river Betwah, some fifty miles from the Sinde. The cholera, which had already begun its decline en route, now disappeared from the army's ranks, confirming Hastings in his view, and that of subsequent reporters on the epidemic, that the cholera derived from local miasmatic causes and that the salvation of the army lay only with its "change of ground and climate."[5] On November 22, Hastings heard the unfamiliar sound of laughter in the camp and breathed a sigh, knowing the crisis had passed. The British Army in India had narrowly escaped total destruction at the hands of this new, fearsome, "epidemical" cholera.

FIGURE 4.2. *A View of Erich above the River Betwah*, as depicted by an anonymous artist attached to Hastings's expeditionary force in 1817. The cliffs about the Betwah provided a safe haven for British troops following the cholera outbreak. The river is considered sacred by tradition and is mentioned in the Sanskrit epic the *Mahabharata*. (© British Library, Marquess of Hastings Collection.)

Percy Shelley's literary-minded cousin, Thomas Medwin, served as an officer in Hastings's army in that fateful campaign. His experience along the banks of the Sinde "haunted" him the rest of his life:

> One march I shall never forget . . . I was in the rear-guard, and did not get to my new ground till night, and then left eight hundred men, at least, dead and dying, on the road. Such a scene of horror was perhaps never witnessed. . . . We lost a whole troop.[6]

Sometime after that traumatic experience, perusing a street bookstall in Bombay, Medwin came across a volume of Shelley's poems and resolved to renew his contact with his young relation, in whom he discerned the marks of poetic genius. After his discharge from the army, he joined the Shelleys at their new expatriate home in Pisa in October 1820.[7]

One evening at the Casa Galetti in Pisa, when more high-minded themes were exhausted, Medwin turned the conversation to his terrifying experience with Hastings's army in 1817. It must have seemed, to veteran members of the Shelley Circle, like a return to the creepy horror stories of the summer of 1816 in Geneva. We can imagine their stricken response to Medwin's tale of a vast bustling camp reduced to the shocked silence of a morgue, healthy soldiers collapsing midsentence in their tents, and the nightly bonfire parades of Indian camp followers bearing their dead to the river. These stories of the Indian cholera made Claire Clairmont's blood run cold, and she reflected harshly on Medwin's conversational taste in her journal: "A bloody war, a sickly season, a field officer's corpse."[8]

Cholera, alas, was never a sentimental disease well adapted for parlor conversation. When, that same year, the Shelleys' friend John Keats succumbed to tuberculosis in Rome, the poet's early death inspired Shelley's elegiac masterpiece "Adonais." Four years later—when Shelley himself was already dead—Lord Byron fell victim to a malarial fever in Greece, having joined the armed cause for independence. That was perhaps the definitive Romantic demise. Death by tuberculosis or malaria was well enough, but cholera, the most feared and written-about nineteenth-century disease, neither communicated an aura of wistful decline (Keats) nor sanctified the soul-endowed sufferer with tormenting fevers (Byron). Instead, it was utterly dehumanizing. In minutes, cholera turned a walking, talking person into a sluice. Microbial agents seized the body, drowned it, and drained its life-giving fluids before abandoning the corpse in its own waste. Romanticize that! The Marathan word for the cholera, *moredesheen*, was Frenchified by Europeans in India as *mort de chien*: to die like a dog. Medwin had brought the horrors of cholera home to the Shelleys for the first time (but not the last), and his ghastly tale must have made for a rare awkward silence at the Casa Galetti.

That night in Pisa in 1820, the deaths of Keats, Shelley, and Byron all lay in the future. None of the company thought death was so near, just as they could not have imagined the geophysical teleconnections between the wet, stormy summer of 1816 they had spent together in

Geneva and the chilling war stories of cousin Thomas, returned all the way from India. Spellbound by cholera's aura of a modern-day plague, Mount Tambora, if they had heard of it, could not have been further from their thoughts. Our situation is different, however. With the aid of recent, groundbreaking research into the dynamics of cholera and climate, Tambora will loom large in the history of the nineteenth century's greatest killer.

THE YEAR WITHOUT A MONSOON

Beginning immediately after its eruption on April 10, 1815, Tambora's volcanic dust veil, serene and massive above the clouds, began its westward drift aloft the winds of the upper atmosphere. Its airy passage to India outran the thousands of waterborne vessels below bent upon an identical course, breasting the trade winds from the resource-rich East Indies to commercial ports in the Indian Ocean. The vanguard of Tambora's stratospheric plume arrived over the Bay of Bengal within days. In late April in Madras, on the southeast coast of India, the morning temperatures plummeted over the course of a single week from 50° to 26°F, a harbinger of Tambora's profoundly disruptive effects on the weather systems of India.[9]

A 1999 study of ash deposits in the Arabian Sea off the coast of modern-day Pakistan indicates a heavy aerosol load in the subcontinental atmosphere in the aftermath of Tambora's eruption.[10] By the time of its arrival in the Indian subcontinent, however, the Tambora cloud had already lost the greater part of its initial volume. Of the vast mass of rock and ash blasted into the upper atmosphere on April 10 and 11, most particles were too large to remain airborne, cascading gently into the tropospheric realm of clouds and weather—there to be washed out of the sky by rain. This great husk of matter shed, Tambora's sleek, residual cloud—made up of pulverized mineral matter, gases, and sulfate aerosol particles less than a micron in thickness—would hang at altitude for more than two years.[11]

Tambora's impacts on the Indian monsoon were not immediate. The sheer presence of a large volcanic cloud in the stratosphere is less important than the sequence of chemical reactions it sets in train. Released from their eon's residence in the earth, Tambora's sulfate gases embraced the freedom of the oxygen-rich sky by forming new molecular combinations: first sulfur dioxide then, with increasing oxidization and interaction with water vapor, tiny droplets of sulfuric acid. As the months passed, the volcanic aerosols, further plumped by water vapor drawn to their acidic content, reached a crucial tipping point: around the end of 1815, Tambora's aerosols attained a density sufficient to interact with both the sun's rays *and* the radiative heat from the earth, reflecting incoming solar energy back to space (the albedo effect), while at the same time intercepting longwave radiation from the surface.

The sum of these effects was a hotter stratosphere but a net cooling of surface temperatures, initiating a three-year depression of the thermal cycle of the South Asian continent, and ultimately the globe. This depression of summer minima and maxima—at the height of the growing season—proved devastating to farmers in the temperate zones of the North Atlantic in 1816 and 1817. But in the tropical latitudes of South and East Asia, raw surface temperature decline was less important than the impact of a disrupted thermal synchrony between land, sea, and sky on the life-giving monsoon.

The ecology of the Bengali river delta is inseparable from its monsoonal climate. From their first encounter with the South Asian continent, European travelers identified the monsoon as its defining cultural and economic driver. Dry through much of the year, the land would be uninhabitable but for the awesome three-month deluge that replenished aquifers and generated crops. In addition, monsoonal winds brought trade. For centuries before the British arrival in the 1700s, Arab merchants—and later the Portuguese and Dutch—had sailed the monsoonal courses to India and Africa, where they bought spices and slaves to sell on the Mediterranean market.

In the dry, cooler months from November to March, the prevailing winds in India come from the north. Then in May, the geophysical

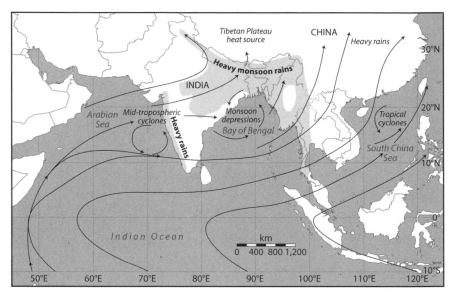

FIGURE 4.3. A synoptic map showing the major weather systems associated with monsoonal fluctuations across South Asia. In summer, a low-pressure system over land results in the influx of moist air from the Bay of Bengal, while in the winter dry season the reverse occurs: a high-pressure system drives moisture away from the landmass. (Bin Wang, ed., *The Asian Monsoon* [Berlin: Springer, 2006], 145.)

mechanisms of the Indian wet monsoon begin to stir.[12] As the Earth's summer tilt draws the sun's maximum heat to the latitudes just north of the equator, the land responds more quickly to the increased solar input than does the vast thermal sink of the Indian Ocean. The temperature gradient between land and ocean steepens, awakening the cloud-bearing monsoonal winds from the south, which rush en masse into the depressurized atmosphere over the Indian continent. The winds abruptly swerve northeast, according to the rotation of the Earth, and head directly for the Bengal delta, bearing storm clouds that brood and toil over the Indo-Gangetic plain for weeks on end, powered by an atmosphere crackling with energy. Cooling as it rises over the heated earth, the moisture condenses, depositing wave after wave of torrential rains onto the parched earth in the form of cataclysmic storms. At the height of this annual meteorological drama, the Ganges River takes to imitation of its ocean neighbor: gale winds from the south turn the river current against itself, propelling storm surges over the stone *ghauts*, bringing down houses built too daringly near the banks, while river traders moor their craft to the shore, hoping to survive long enough to enjoy the blessings of the *sharif*, the monsoon crop.

But in the Tambora year 1816, these rains wouldn't come.[13] As long as the volcano's sulfate veil cooled the Earth, inhibiting evaporation from the ocean and deflating the temperature differentiation of land and sea, the Indian monsoon lacked its vital motive fuel. In a recent climate modeling study that examined the impact of tropical volcanoes on the Asian monsoons, scientists from the National Center for Atmospheric Research found "significant" alteration in the hydrological cycle of the monsoon in the aftermath of a high-impact eruption, with the strong likelihood of "large reductions" in summer rainfall in South Asia.[14]

Across the breadth of the Indian Ocean, the trade winds faded, stalling thousands of ships on their westward course. Farmers on the Ganges plains waited in vain for the expected shift of winds to the south bearing ocean-fed storm clouds. For the vital *sharif*, timing was everything. But when to plant? No chorus of birds gave the familiar sign, levitating above the trees like flags before the arriving wind. In the absence of the seasonal rains, essential ecosystem services for the human population of the Ganges delta deteriorated rapidly. Because of the inwelling tides from the bay, the river waters were considered poor for drinking. Ordinary people relied on rainwater ponds, called "tanks," while the affluent drew upon sweetwater artesian wells. In the absence of monsoonal replenishment, tanks and wells alike grew fetid. For desperate Bengali villagers, the failure of the monsoon represented a divine judgment. They reflected now on what terrible sins had been committed in their communities for the monsoon to forsake them so completely. Through the middle months of 1816, prayers and *pujah* ceremonies rang through Hindu temples all along the Ganges. The holy men consulted the stars or divined the intent of the winds from the smoke of coconut oil burned in its shell.

During this driest of Indian Mays, wild fluctuations between excessive cold and record heat saw Bengalis and Europeans alike "drop down dead in the streets."[15] This regional South Asian heat wave, in an era of general volcanic cooling, may be explained by an enhanced meridional circulation. In the northern hemisphere, Tambora's stratospheric impact prompted an outflow of Arctic air to the temperate zones of Europe and New England, cooling the landmasses. Meanwhile, on the subcontinent of India, a counterbalancing circulation, possibly allied with a cyclical El Niño system, produced a northerly poleward flow of tropical air through Bengal, spawning localized heat spikes and drought. For the suffering Bengalis, the June rains were inexplicably scanty. Vital tributary streams across the delta dried up, threatening the rice crop. This crippling monsoonal break, lasting until late August at least, is the longest in the historical record of the Asian subcontinent. A tree-ring study of Himalayan cedars from 2007 shows an "extreme low growth" in trees all across the river basins of northern India in 1816, an indication of severe moisture stress.[16]

When, too late, the volcanic retardants in the atmosphere were overcome and the depressed monsoonal machine returned to furious life, the now unseasonal rains it carried were ruinously extreme. The drought of 1816 subsequently gave way to hundred-year floods, bringing a second season of failed crops, famine, and misery to Bengal. In September, typically a month of monsoonal decline, monster storms inundated the delta on a scale remarkable even in a region accustomed to seventy inches of seasonal rain. Various local disease outbreaks—designated "Bilious Fever" or simply "Malignant Sore Throat" by perplexed British doctors—took hundreds of victims. In the 59th Regiment stationed in Jessore, north of Calcutta, as many as a dozen soldiers perished daily, while the banks of the rivers across the Ganges delta were "covered at all times with the dead and the dying" from the local villages.[17] All this, however, was but an overture to the main act of the Tambora climate emergency in India.

DEATH ON THE GHAUTS

As bad as 1816 had been, 1817 in Bengal began like no other year. January and February were looked forward to as months of serene weather:

cold, clear nights and morning fogs rising like a curtain onto sunny days freshened by breezes from the great mountains to the north. But Tambora's second year announced itself instead with clouds and heavy rain. The winds veered crazily from the north to the east, then to the south, bringing drenching downpours. On March 21, an unprecedented hailstorm destroyed the spring grain crop and tore up orchards of dates, bananas, and papaya all across the fragile alluvial plain.

The disease cholera had been seasonally *endemic* to Lower Bengal since time immemorial. Traveling the narrow coastal roads of the Sunderbans in the Bay of Bengal today, one still comes across shrines dedicated to Ola Bibi, the goddess of cholera.[18] But in May 1817, when the monsoon arrived three weeks early and delivered further calamitous quantities of rain to the delta region, the perennial Bengal cholera suddenly appeared out of season, showing unusual strength and breadth. By August an unprecedented "epidemic" of the disease had spread among the Indian population. A short month later an official report declared cholera to be "raging with extreme violence" through both the Indian and European populations of Bengal but also embarking on an unprecedented outward course to the north and west, following the river.[19]

For the early nineteenth-century European tourist, the uncultivated riverbanks of the Ganges offered the gratifying sight of monkeys, buffalo, and occasional elephants wallowing in the mud. On the river itself water-loving lotus flowers and lilies brushed along the boat's hull, signs of the nutrient-rich ecology of the delta. "A paradise of flowers," was one Englishwoman's commentary on her pleasure trip from Calcutta to Benares.[20] At a bend in the river, the source of these botanical offerings might hove into view. The local *ghaut*—broad stone steps leading directly to the river—served as a kind of Indian town square and impressed foreigners as a grand forum of daily village life. From first light, people could be seen washing clothes and bathing, while women bore jars of water on their heads back up the steps. At the summit of the *ghaut* invariably stood a temple, where dedications were made to the sacred river on which all these essential activities depended.

In 1817, this vivid, public life of the *ghauts* was transformed by cholera into a parade of horrors. All along the Ganges River, the busy tempo of village life gave way to mourning and public immolation of the dead. In the eyes of one English traveler, a clergyman named James Statham, the cholera epidemic transformed the vibrant and picturesque river life of the Ganges into a scene from Dante's Hell:

> None but those who have witnessed the distressing sight can form an adequate picture of human misery which the *ghauts* afford at the time when the cholera rages. The dead and dying are all huddled together in a confused mass, and several fires are blazing at the same time, consuming the bodies of the more rich and noble, who have just died, whilst the poor creatures who are expiring feel certain that in a few minutes their bodies must share the same fate, or be hurled into the flowing stream, to become the prey of waiting alligators, or, what is worse, to be left on the beach, a prey to jackals and vultures, which infest the spot. Fresh arrivals every hour multiply the misery, as groans and cries increase, while the stench proceeding from the burning bodies, and the lurid gleams of the blazing fires reflected by the water, and giving somewhat of an unearthly appearance to the features of the suffering victims around, furnish a scene of woe which completely baffles the power of description to portray.[21]

An almost hysterical fear of this new, insatiable cholera—inspired by its shocking symptoms and territorial expansion—seeps through the restrained bureaucratic prose of British colonial officials. The epidemical "cholera morbus" was an "evil," "a horrid scourge," "an awful and desolating calamity" that "threatened to sweep off a large portion of the Native population" if some means could not be found to counteract it. It was "a malady more destructive in its effects, and more extensive in its influence than any other recorded in the annals of the country." As a consequence of this high mortality and general panic, the cholera threatened social order, the proper running of the Indian economy, and thus company profits. "Much mischief has arisen," complained the Board of Revenue in Calcutta, "from the great alarm of the people, their quitting their habitations and their proper pursuits."[22] Death

FIGURE 4.4. William Hodges, *The Ghauts at Benares* (1787). (Royal Academy of Arts, London; Photo: John Hammond.)

counts were hazy, perhaps inflated, and suspiciously round. Certainly thousands, perhaps tens of thousands, died in the first season. By the 1830s, at the beginning of the global cholera panic, European estimates of Indian fatalities since 1817 would run into the millions, numbers impossible to verify.[23]

CHOLERA AND CLIMATE CHANGE

Endemic cholera in Bengal had traditionally been associated with the "winter" months of November to January, with a smaller peak in the hot, dry months of April and May. The reach of the disease remained limited, in any given year, for the simple reason that it soon ran out of fresh victims. Explanations for the unprecedented epidemic outbreak in Bengal in 1817 depend upon the putative emergence of a new strain of cholera capable of bypassing the built-up immunity of its indigenous hosts, then spreading rapidly to successive populations in various directions.[24]

Energized by its passage through the human intestine, the cholera microbe achieves a temporary hyperinfective state. Density of human traffic is thus essential to its continued transmission. Carrying the fatal microbe with them in their bowels, soldiers, pilgrims, and traveling merchants in 1817 dispersed the infection to unsuspecting fresh host communities north and west across India. In the following years and decades, these same human vectors, following the webs of global trade, brought cholera southeast to the Dutch Indies and East Asia, and northwest across the great trading routes of Arabia to Russia, Europe, and finally the Americas. The eventual global death toll of nineteenth-century cholera stands in the tens of millions.

The "father of British medical writers on cholera," Calcutta physician James Jameson, traced the cause of the 1817 cholera to abnormalities in the Bengal climate in the two-year period leading to the outbreak. His classic 1820 report to the Calcutta Medical Board includes a ninety-page prefatory description of the "distempered" state of the weather beginning in late 1815. By attributing the outbreak of cholera to meteorological causes, Jameson placed himself within an established tradition of environmentalist medical theories dating back to the Hippocratic revival of the late seventeenth century. The pathogens of disease, whether produced by the noxious exhalations of the Earth or the "vitiated" state of the atmosphere, were airborne. The Indian cholera, Jameson concluded, had been spread by the strangely humid atmosphere, drought, and unusual winds of 1816–17. With its detailed synoptic data—based on one hundred survey reports from physicians across British India—Jameson's report is a landmark in medical and public health literature and was the most quoted text by English writers on cholera for decades.

Already by the Tambora period, however, debates over the etiology of cholera were shifting. In subsequent decades, a new legion of medical theorists of cholera would come to reject Jameson's meteorological emphasis in favor of an emerging liberal paradigm that captured the imaginations of progressive physicians and public officials through the Victorian age. Infectious disease was not "natural" but rather the product of human-created filth, of the open sewers and fetid air of slums and industrial tenements. Cholera was a social disease, an index of failures by nation-states to regulate and sanitize their colonial ports of trade and booming industrialized cities. In the pulpits of Europe, the cholera would accordingly change shape from a divine punishment for wickedness to a progressive moral calling for social reform, to provide hygienic living conditions for the newly urbanized masses.

Thus the early climatological theories of cholera rapidly lost ground during the heroic age of nineteenth-century sanitarianism. For almost a century after the apparently definitive discovery of the comma-tailed cholera bacterium in a Calcutta pond by pioneer bacteriologist Robert Koch in 1883, both clinical theory and public health policy surrounding the disease endorsed the emerging contagionist consensus, focusing on its human-to-human transmission. The fecal matter of infected persons in waterways was simply the pathogen's mode of transit from one human host to another, which might be denied by proper investment in sanitary engineering. This dominant model of twentieth-century cholera science relegated Jameson—and a host of other pre-Victorian writers in the medico-meteorological tradition—to the dustbin of history, as embarrassing examples of the puffed-up guesswork and shamanistic fantasies that passed for medical science before the discovery of bacterial infection. Jameson's enlightenment medical geography, with its environmentalist perspective on disease transmission, had lost the battle of ideas, which resolved by century's end into a narrower medical-scientific practice focused on the career of pathogens, to be charted by the new techniques of laboratory biology.

Now, however, at the beginning of the twenty-first century, the medico-meteorological worldview of James Jameson has experienced a second coming. The post-Victorian bacteriological consensus on cholera has been overtaken by a new, more complex etiological paradigm that restores credibility to the early nineteenth-century model of climatic disease dynamics. Beginning in the late 1960s, epidemiologists utilizing the tools of modern molecular biology discovered the *vibrio cholerae* thriving in nonendemic form among the zooplankton and protozoa of a wide range of aquatic environments—from the Chesapeake Bay to the lochs of Scotland—independent of human hosts. Not the "Asiatic" cholera after all! Nor are humans the end point or object of the cholera. We are merely accidental part-time hosts to a pathogen that enjoys ancient privileges in a range of aquasystems outside the human intestine. From its ancestral origins in the deep sea, the *v. cholerae* resides year-round in brackish reservoirs around the globe, while its pathogenic strains are favored only within dense human communities situated at low elevation along estuarine coasts, in tropical climates characterized by high temperatures, humidity, and heavy seasonal rain.[25]

To succeed in this global role, the cholera bacterium possesses an unusually flexible and adaptive genetic structure highly sensitive to changes in its aquatic environment. In the monsoonal waters into which the five major rivers of the Ganges delta flow, the *vibrio cholerae* prospers by attachment to benign organic hosts—plankton, algae, crustaceans, and even tiny insects—where they participate in the mineralization of matter vital for the replenishment of the aquatic food web. The circulation patterns and organic life cycles of the northern Indian Ocean are unique on account of the monsoonal climatic regime. Over the Bay of Bengal in June and July, the water vapor content of the atmosphere reaches its highest level anywhere in the world. The prevailing winds reverse twice annually, churning up the waters of the bay, while the summer rains create a vast freshwater runoff from the rivers. Mile-wide phytoplankton blooms, on which great colonies of *v. cholerae* depend, surface at the river mouths and glide along the coast by the East India coastal current according to a distinct but variable seasonal pattern that drives the bay's dynamic interaction with the cholera microbe.

Answers to the question of why a new and deadly cholera strain developed in Bengal in 1817 had been mostly speculative until the complete sequencing of the cholera genome was published in 2000.[26] Since then, investigations of cholera genetics have drawn the outline of an evolutionary narrative for cholera, by which the post-1817 pathogenic strains of the bacteria separated themselves from their benign or strictly endemic marine ancestors.

In a series of articles, University of Michigan ecologist Mercedes Pascual has explicitly revisited the early nineteenth-century debate between meteorological and contagionist theories of cholera, proposing to "integrate" them and thereby return medical science full circle to an understanding of "the influence of climate on disease dynamics."[27] Pascual's theoretical ecology of cholera has shown "interannual variability" of climate—that is, weather anomalies such as drought, flood, and unseasonable temperatures—to be a strong driver of outbreaks. In addition to threatening the security of water infrastructure, excessive rainfall alters the salinity levels of water and promotes the growth of nutrients conducive to bacterial production. Conversely droughts, by increasing the temperature of reduced standing bodies of water and concentrating the bacterial population, also promote disease transmission.

Fellow cholera epidemiologist Rita Colwell has likewise singled out extreme meteorological events, with their impacts on water temperature, salinity, and conditions of flood or drought, as capable of both amplifying transmission of cholera and producing the nonlinear transformation of organic pathogens into new and potentially deadly forms. For the new generation of cholera epidemiologists, such as Pascual and Colwell, the trigger for cholera is thus climate, specifically climate *change*. Lateral genetic transfer, by which the cholera bacterium is modified by foreign elements, is promoted by changes in environment—in the temperature, salinity, and alkalinity of the aquatic habitat.

If a cyclical meteorological event such as an El Niño is capable of amplifying cholera conditions, as Pascual has shown, then a truly bizarre monsoonal anomaly such as the Tamboran Asian weather regime of 1815–18 would certainly have been sufficient for the evolution of a new microbial strain. In 1817, the aquatic environment of the Bay of Bengal had deteriorated radically owing to the disrupted monsoon, a consequence of Tambora's dimming presence in the stratosphere. By a process that remains mysterious in its details, the altered estuarine ecology then stimulated an unprecedented event of genetic mutation in the ancient career of the cholera bacterium.

From this point the cholera's path is chillingly clear. Changes in water temperature and salinity promote the bloom of zooplankton, the cholera's main aquatic host, while flooding dredges up deep-welling nutrients and transports the pathogen into the water system of coastal human communities. Barely above sea level, the waterways of the Bengal delta ebb and flow with the tides of the bay. If changes in rainfall patterns drive sea levels higher, bay waters ooze and swell inland, infiltrating the ponds and tanks dedicated to human use. The mix of fresh and salt water keeps the plankton blooms, and their bacterial tenants, at the surface. From there, the *v. cholerae* is but a cup of water, rice pot, or shellfish meal away from colonizing its first human victim. And if that first host is tended to in unsanitary conditions and his waste finds its way back to the water sources of the community, the stage is set for cholera to rise up from the waters in its ghoulish mortal shape.

In short, cholera outbreaks are climate driven, and cholera is a climate change disease. In our Tambora case study, droughts *and* floods—twin climatic extremes characteristic of the 1815–18 period in Bengal—provided hospitable conditions for the emergence of a new epidemic strain of cholera, with devastating global consequences. The sustained weather anomalies in Tambora's wake impacted, by the physical laws of teleconnection, molecular processes in faraway Bay of Bengal. The long-forgotten "father of cholera science," James Jameson, was thus right to contemplate the distempered skies over Bengal when charged with reporting on the 1817 cholera epidemic for the Calcutta Medical Board. Because of his meticulous record keeping, we are able to reconstruct the deteriorated monsoonal conditions in South Asia in the aftermath of Tambora. And with India's Year without a Monsoon brought into focus, the vision of a nineteenth-century world shaped by climate-driven epidemic cholera emerges for the first time. Tambora's Frankenstein weather, wild and weird, created a microbial time bomb in the waters of the estuarine Bengal delta. Once exploded, life on Earth, at least for human beings, became a far more dangerous proposition.

CHOLERA GOES GLOBAL

In the years after the 1817 epidemic, of course, none of this was known, which made the mysterious new global scourge all the more terrify-

FIGURE 4.5. In the mid-1820s, Scottish artist James Baillie Fraser produced a series of Indian sketches titled *Views of Calcutta and Its Environs*. The high quality and expense of the volume—a nineteenth-century "coffee table book"—indicates a considerable market for Indian-themed art in Britain around the time of Tambora. Here is Fraser's impression of the grand bazaar in Calcutta, which presented a very different scene during the devastating cholera epidemic a few years earlier. (Yale Center for British Art, Paul Mellon Collection.)

ing. The disease soon confounded the Jameson school of medicometerological scientists, since cholera declared itself at home across multiple latitudes and climatic zones. In 1819–20, the epidemic retraced the path of Tambora's volcanic plume southeast across Burma and Siam (Thailand), where cholera's victims clogged the rivers and canals of Bangkok, plunging the kingdom into crisis. From there the epidemic spread to Java, where Tambora's curse performed a kind of tragic second coming, killing an estimated 125,000 people, more than died in the volcanic eruption itself.

Mortality throughout the East Indies was catastrophically high. Hitching a ride along the routes of trade, the cholera then spread north to the Philippines, Japan, and China, westward over land and sea to Persia in 1822, and from there by caravanserai to the southern borders of Russia. Medical historians disagree on what happened next: whether

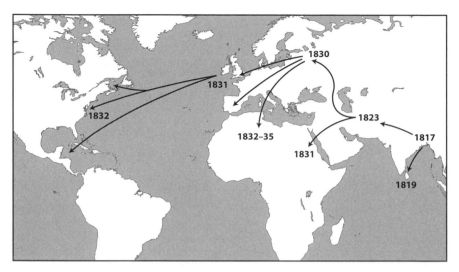

FIGURE 4.6. Map showing the global spread of epidemic cholera out of Bengal post-1816, in what has since been termed the first cholera "pandemic." Epidemiologists have identified six subsequent pandemics, the most recent, and ongoing, originating in Indonesia in 1961. (Adapted from John Barber, *An Account of the Rise and Progress of the Indian Spasmodic Cholera . . . Illustrated by a Map Showing the Route and Progress of the Disease from Jessore, Near the Ganges, in 1817, to Great Britain, in 1831* [New Haven, 1832].)

the 1817 cholera merely slowed before gaining renewed strength in 1829 when it reached Moscow, or whether a new epidemic began in Bengal in 1826, reaching the Russian capital three years later. Either way, on an implacable westward march that inspired as much fear in those awaiting its coming as misery in those already afflicted, cholera reached Paris in 1830, England in 1831, and finally American shores the following year.

The eerie predictability of cholera's progress, because of its perfect alignment with the global web of human commerce, "created panic on a scale not seen since the great plague epidemics of the seventeenth century."[28] For all the anticipation, however, cholera's first strike was always shockingly sudden. In April 1832, the poet and journalist Heinrich Heine wrote about his experience at a masquerade ball in Paris at which a guest, dressed as a harlequin, suddenly collapsed on the dance floor. Panic ensued. Dozens of other victims fell to the floor until the ballroom resembled a battlefield. These fresh cholera victims, who had begun their evening dancing at a carnival ball, ended it rudely tossed into makeshift graves still dressed in their masquerade costumes.

The fact of cholera's essential mystery and resistance to treatment only amplified its power over the nineteenth-century imagination. An extraordinary number of speeches were given, orders issued, and weighty words written giving voice to these universal anxieties. Cholera is accordingly the best-reported disease in history and boasts a vast medical and popular literature. In the twentieth century only AIDS approaches cholera in terms of the volume of research and public opinion produced. As part of this same cultural phenomenon, cholera was the first disease to come under modern public health surveillance and gave rise to entire new bureaucracies across the European nation-states. Cholera operated, historian Christopher Hamlin has suggested, "like the conscience of the nineteenth century," drawing the plight of the newly urbanized European poor into stark focus and helping bring into being the modern apparatus of professionalized public health.[29] The first British Board of Health was established in 1831 in response to the cholera epidemic, while in 1851 the first International Sanitary Conference was held in Paris—cholera its sole subject. By century's end, the modern sciences of microbiology and epidemiology had formed around the iconic image of a white-coated cholera scientist peering through his microscope in a laboratory, while sanitarianism assumed its status as the global orthodoxy of disease prevention, which it still holds today.

The worldwide cholera epidemic originating in Bengal in 1817 also changed the character of European colonialism and global race relations. Writings on the impact of climate on human physiology and character post-1817 show a marked shift from an optimistic theory of acclimatization—in which even European skin color might naturally be expected to change—to a rigid hereditarian model, whereby distinct races were tied to their respective environments by virtue of their indigeneity—that is, their long-term, continuous habitation of a place. History, according to this new logic, became biology, and eighteenth-century environmental determinism began its insidious evolution toward nineteenth-century biological theories of race. The susceptibility of Britons to tropical disease, horrifyingly exposed as never before in 1817–18, and then later in the European cholera panic of the 1830s, helped consolidate views of India as a dangerous, alien environment to which Europeans could never adapt.

According to the simple, linear model of disease transmission dominant in the nineteenth century, mobile human populations were the exclusive carriers of cholera. Indians naturally blamed the British army and merchants for spreading the disease, but Europeans chose religious pilgrimages as the principal culprit. In India, the seasonal traffic of thousands toward sacred rivers captured the suicidal folly of Eastern idolatrous practices. Closer to home, and thus more politically inflammatory, was the pilgrimage to Mecca, site of dozens of cholera outbreaks through the nineteenth century beginning in 1831 and the source of rabidly anti-Muslim editorializing in Europe and North America. Perversely, then, despite its global presence and indiscriminate path, cholera spawned ways of thinking about the *differences* between East and West. For Europeans, the racialization of cholera converted an inexplicable and frightening cultural threat into a blame narrative with a readily identifiable human enemy. As such, cholera discourse became a cornerstone of Western orientalism, the racist legacies of which continue to pervade geopolitics in the twenty-first century.

It is thus difficult to overstate the impact of nineteenth-century epidemic cholera—a deadly product of Tambora-era climate change—on world history. The disease was "an insult to progress" itself, undermining bourgeois confidence across Europe, North America, and the colonial dominions by exposing both the vulnerability of the new global marketplace to disease and the gross economic inequalities evident in the slums that harbored it.[30] The connection between cholera and class struggle is borne out in the revolution-rich dates coinciding with major outbreaks—1831–32, 1848, 1871, and so forth. The Tambora-driven cholera of the middle 1800s thus echoes the cataclysmic Black Death event of the fourteenth century, which was marked by riots and civil uprising. Cholera was the Victorian age's worst nightmare. Its symptoms exposed the excretory functions of the private body in the most humiliating way imaginable, while its victimizing the poor and oppressed exposed the thin veneer of polite, prosperous modern society. Just as the abolition debate politicized race, cholera politicized class, so that sanitation emerged with slavery as the premier social justice issues of the nineteenth century.

THE LAST MAN

And yet there is no great Victorian novel about cholera. This is surprising, at first, given the centrality of cholera to nineteenth-century social history. But novels are Romantic forms of art, in which empathic young heroes and heroines flower into self-consciousness and emerge as moral actors over time and hundreds of pages. Cholera was simply too brutal and disgusting a death for these narratives—even for punishment of the wicked. Cholera does not make stories; it ends them. Imagine Jane Bennet contracting cholera instead of a head cold at the beginning of Austen's *Pride and Prejudice* (1813) and infecting all of Meryton. No Darcy, no Elizabeth, no delights of Pemberley, just corpses in soiled sheets rolled up as terror grips Hertfordshire.

But if there is a nineteenth-century cholera novel, it is Mary Shelley's *The Last Man* (1826). Shelley's post-apocalyptic description of a world depopulated by epidemic disease, with a tiny rump human community meditating on the end of history, is more existential than romantic in character. This may account for the novel's lack of popularity, especially when compared to the iconic *Frankenstein*. But in the aftermath of her own cascading personal tragedies in the period 1818–21—the death of three children, followed by the soul-emptying loss of her husband, Percy Shelley—Mary turned to descriptions of cholera coming out of India for her new novel and to the morbid tallies of its victims. In the novel, she locates the origin of the mysterious plague, as did James Jameson, in "perturbations" of the atmosphere. Cholera, a phantom agent of death that was brutal, unknowable, and potentially limitless in its reach, seemed to map her personal experience of mortality onto a vulnerable world.

However therapeutic the writing of *The Last Man* might have been for her, Mary Shelley found, as the scientific community has done, that cholera does not readily submit to description or catharsis. Just as the *v. cholerae* is a permanent cohabitant with us on our watery planet, so one can never be "done," in mind or art, with cholera. Six years after her publication of *The Last Man*, Mary Shelley's half brother, William, became one of the last cholera victims in London's 1832 epidemic—his death a late installment in her lifelong litany of sorrows. William Godwin died of cholera in a polluted neighborhood of London. That much Mary could have known. But for us, the full cause of his death may only be properly measured across far wider vectors of time and space: from the fiery plume of the exploding Tambora in 1815, to the salty estuaries of the Bengal delta, to the humid intestines of merchants traveling the myriad trade routes of the nineteenth-century world in their ships and caravans. By the far narrower human accounting of literary history, not one but two of Mary Shelley's novels may now be traced to Tambora. As a meteorological event in the summer of 1816, it helped inspire the celebrated *Frankenstein*. But it no less set the ecological conditions for Shelley's *The Last Man*, a tale of global devastation by climate-driven disease.

The last word goes to the Bengalis, who have suffered the tragedy of cholera more than anyone. In one local cholera legend, a man travels on business from Calcutta to a village along the Ganges. He is welcomed into the house of his associate, where he dines and is entertained. He is surprised at the absence of servants—for who has cooked the meal? And when he requests a candle to lead the way to his sleeping quarters, it is fetched for him instantly without his hosts appearing to leave the room. Undressing for bed, the guest looks out the window, where he glimpses the dim figures of the man and his wife pass into the garden on a side of the house with no door. Consumed with terror, he is unable to sleep. At first light, he runs to the street. But there is nobody about. There it dawns on him that he has spent the night in a ghost town. All have died of the cholera, and the village is silent and empty.—He is the last man.

CHAPTER FIVE

THE SEVEN SORROWS
OF YUNNAN

We know from a British emissary onboard a ship bound for Canton in the summer of 1816 that Chinese skies still bore their lurid volcanic imprint some fifteen months after Tambora's eruption. The description is among the best we have of the most remarkable volcanic weather seen on Earth for perhaps thousands of years:

> The evening of the 9th of July, the sky exhibited such novel though brilliant appearances, as led us to fear that they would be followed by formidable changes of weather. The course of the sun, as it sunk beneath the horizon was marked by a vivid glory expanding into paths of light of the most beautiful hues. They did not in the least resemble the pencils of rays which are often seen streaking in the sky at sunset, but were composed of sheets of glowing pink, which diverging at equal distances from the sun's disk, darted upwards from the horizon, diminishing in intensity of colour, till they vanished in the azure of the surrounding atmosphere.[1]

This much can be said of Tambora's plume: it was an attractive killer. A tragedy of nations masquerading as a spectacular sunset.

One of the many bureaucratic and intellectual achievements of China's two-thousand-year empire is its meteorological record keeping, which surpassed that of any other nation in historical reach and detail. The fact that climate was reckoned so closely with crop yields means that, in reconstructing the global impact of a large-scale climate event such as the Tambora disaster, China is a fruitful place to turn for close-grained, regional analysis of its environmental and social effects.[2]

We know, for example, that the tropical latitude of the Chinese island of Hainan did not protect it from summer snows in the summer of 1815 or a severe winter in which more than half the forests perished. Eastern China likewise suffered from a suite of record low temperatures and crop failures in the Tambora period. In Shanxi, the crop-killing summer frosts of 1817 heralded mass emigration from the province, reminiscent of the large-scale refugeeism being played out across Europe and New England at the same time.

But no region in China, it appears, suffered so greatly as the southwest province of Yunnan.[3] When record low temperatures persisted with unprecedented severity over three successive growing seasons, prosperous Yunnan descended into a Confucian nightmare in which all sacred bonds of community were severed. Famished corpses lay unmourned on the roads; mothers sold their children or killed them out of mercy; and human skeletons wandered the fields, feeding on white clay.

SOUTH OF THE CLOUD

By the time of Tambora's eruption in April 1815, the Chinese empire had achieved the massive territorial dimensions it still commands today, some twelve million square kilometers. One of the last territories to be brought under the Chinese banner, the mountainous province of Yunnan in the southwest had long been integrated within the Indian Ocean trade zone and enjoyed close cultural ties to the lands of the Mekong delta to the south: modern-day Burma, Laos, and Vietnam. A key transit point along the ancient silk road, Yunnan has served as a crossroads of cultures for millennia.[4] When the legendary Mongol emperor Kubla Khan sent Marco Polo to the capital Kunming in the 1280s, the intrepid Italian marveled at its diversity of peoples and religions. He also noted its abundance of rice and wheat fields, scattered across some two thousand valleys nestled in the mountains. Despite its elevation and rugged topography—the arable land is only 6% of its surface area—Yunnan is highly favorable to agriculture thanks to these fertile, intramontane lowlands, called *bazi*, which range in size from less than a kilometer to hundreds of kilometers across.

Until the disastrous upheavals of the nineteenth century, Yunnan enjoyed a more peaceful history than might be expected given its geographical circumstance, perhaps on account of its equable climate, which is likewise defined by convergence of differences.[5] The extremes of China's monsoonal climate are felt least in Yunnan, which acts as a buffer zone between the spectacular Indian monsoon and the rival Asian monsoon that largely dictates China's weather east of 105°E. The province's famously pleasant climate depends first on topographical discouragement of the southwest monsoon out of the Bay of Bengal. These grandiose storms are drained of their humid rage by their mountainous overland journey and enter Yunnan as no more than a mild, encouraging breeze and scattered rain showers. In a normal summer, the Tibetan plateau to the northwest—the "roof of the world"—acts as an enormous thermal engine, driving heat upward and depressurizing the atmosphere. The scale of this Himalayan warming is sufficient to offset even the effect of Yunnan's lofty elevation. In Kunming, T-shirts are de rigeur at 1,500 meters above sea level. The southeast trade wind, too, an antiphonal presence, never fully yields to the Indian monsoon, ensuring that Yunnan experiences the fewest summer gales of any region in China.

Yunnan is likewise spared the worst of winter's nick. Its mountains, running north-south, act as a vertical shield against the prevailing cold-air masses spilling down from Mongolia. The upward draw of the steep hillsides maintains a stationary high-pressure system, insulating the region like a meteorological version of the "pleasure dome" Coleridge imagined in "Kubla Khan," his opium-soaked fantasy of medieval China

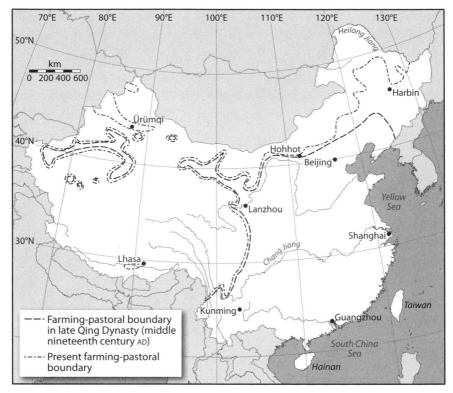

FIGURE 5.1. Map showing the enormous territorial extent of Qing Chinese agriculture, the boundaries of which reached their maximum in the mid-nineteenth century, prior to imperial decline and twentieth-century industrialization. (Zhao Songqiao, *Geography of China* [New York: Wiley, 1994], 50).

written in 1816. Even in January, the highland Yunnanese benefit from the thawing effects of southwesterly winds. Indeed, Kunming enjoys the least variation in diurnal temperature anywhere in China: rarely below 10°C in winter or above 22°C at the height of summer. Awed by this moderation, Chinese folklore has it that there is no summer or winter in Yunnan at all, only spring and autumn.

Yunnan's distinctness from core China is thus reflected in both its history and its climate. The name "Yunnan" means "south of the cloud": from the Chinese point of view this designated a remote and exotic region blessed by perennial warmth. Yunnan's popular moniker, "Land of Eternal Spring," encouraged generations of Han Chinese to venture westward in the boom times of the eighteenth century. Despite its elevation, winters in Yunnan were as much as 4°C warmer than in the flat plain country to the east at the same latitude, owing to its complacent situation beneath a largely stationary southwest air mass. Literally south of the clouds, Yunnan basked in abundant winter sunshine and was thus evergreen, a propitious country for growing rice and other food crops, and with abundant land yet undeveloped.

Until the seventeenth century, the population and agricultural output of Yunnan remained relatively stable. Then in the eighteenth century, at the height of the Qing dynasty's westward march, came a period of enormous growth. The settlement of the west was a remarkable achievement in imperial central planning. That said, the doubling and tripling of Yunnan's population in so short a period—from three million in 1750 to twenty million in 1820—could not have been sustained without an unwritten contract between Qing ambition and a benign climate, without an efficient and easily expandable agricultural system consciously adapted to an environment of pliable soil and sunny weather.[6]

By the turn of the nineteenth century, everything pointed toward the further extension of Chinese control over Yunnan and, more largely, to the province's compliant role in the ever-greater expansion of an integrated, federalized Chinese agro-economic system. The intensification of agriculture across China, wrought by advanced technologies of irrigated land management and fertilization, supported huge increases in population—and was an object of jealous wonder among Western observers, who looked at the comparatively low-yield farming practices in Europe and America with dismay. By 1800 in Yunnan, almost all arable land had been brought under the plow, while more than half-a-million laborers worked in the copper, silver, gold, and salt mines, probably the largest mining operation in the world at that time.[7]

As this thumbnail description suggests, Yunnan resembled many of the world's frontier regions of the early nineteenth century, where long-term indigenous residents gradually gave way to invading settler societies bent on mineral extraction, plowing the land, and building ever-expanding cities as hubs of trade. But this imperial growth plan for Yunnan did not reckon with a period of drastic climate decline at precisely the wrong historical moment. The Tambora-driven weather emergency and ensuing famine of 1815–18 fatally altered the course of Yunnan's development and played its part—as we shall see—in bringing down an empire.

YEARS WITHOUT A SUMMER

Disdained by the filmy Tamboran sun, the Himalayan plateau never warmed during the summer of 1816, nor did the surrounding oceans. In Tibet, just north of Yunnan, it snowed for a remarkable three days in a row in July, where a British surveying party reported famine among the native population and required emergency rations for their own survival. In this altered volcanic summer, the Tibetan plateau reverted to its wintry role, channeling the cold northern air southward and eastward toward the Mekong peninsula. Meanwhile, a cool and subdued Indian Ocean, likewise bereft of its monsoonal power, failed to deliver the warm winds necessary to moderate the cruel Mongolian northerlies.

Yunnan had the misfortune to suffer a cyclical drought in 1814, which meant its grain reserves were already depleted even as its normally benign weather patterns fell under Tambora's chilling spell. The first signs of volcanic weather arrived quickly, a month after the eruption in the spring of 1815. Given the relative coolness of its elevation, Yunnan relied upon unstinting summer sunshine to ripen its grain crops. But in the late spring of 1815, the expected southwest winds did not arrive to disperse the clouds, which instead labored over the mountains, depositing flooding rains that drowned the winter crops. Wheat and barley sprouted underwater, while row after row of broad beans disintegrated into the mud. The bitter rains continued through the disastrous summer and into autumn. The aqueous rice fields might yet have survived had it not been for a frosty August, which strangled the budding rice plants at the critical point of their maturation.

According to one school of agricultural historians, the seven-thousand-year history of domesticated rice production begins in Yun-

FIGURE 5.2. An idealized European impression of Chinese rice agriculture. (Thomas Allom, *China in a Series of Views Displaying the Scenery, Architecture, and Social Habits of That Ancient Empire* [London, 1843–47], 3:26).

nan. At the time of the Tambora eruption in 1815, a fifty-year accelerated settlement program had seen Han Chinese pioneers from the east "reclaim" vast areas of the intramontane lowlands for rice paddies, while picturesque irrigated terraces climbed ever further up the sheer mountainsides. Rice is a famously hardy crop, hence its role as the staple diet for half the world's people. Once a rice-growing system is in place, and refined to allow double or even triple cropping, it will support rapid and continuous population growth, as in the case of eighteenth-century Yunnan. No other plant has the population-carrying capacity of rice, whose system of air passages connecting its roots and stem is formidably efficient, enabling it to self-regulate in widely differing contexts from irrigated fields, to dry upland soils, to riverbeds.

But the cultivation of rice, which is after all a tropical plant, does have its Achilles' heel: cold snaps in summer.[8] A recent study nominated 14°C as the "critical threshold of damage for rice." But sustained temperatures below 20°C, combined with a deficit of sunlight, are enough to spawn the uncontrolled proliferation of reproductive organs within the plant, or otherwise fuse, feminize, or deform them. Instead of ripening into its hardy oval shape, the cold-afflicted rice grain will fail to seal itself and assume the hideous, sterilized form of a tiny spiky claw or twisted stump.[9]

Because of the sustained abnormal cold conditions that prevailed during the Tambora emergency, accounts of the Yunnan famine of 1815–18 are full of descriptions of dehydrated rice grains "withered" or "shriveled" by the much-feared "dark winds" of the north. In East Asia, an old saying has it that a north wind in autumn reduces the rice yield by half. From 1815–18 in Yunnan, the impact of Tambora's cold north winds on rice production may have been more than two-thirds. The branch-like panicles, which in a normal year drooped gratifyingly from the weight of their fruit, never came to flower, their mutant husks empty of grain.

The rice paddies of Yunnan are compensated for the relative coolness of their summer temperatures by the stillness of the intramontane air, allowing the sun to warm the surface of the watery field. When the water temperature is higher than that of the air, the paddy field emits a layer of protective warmth above the plants, insulating them from the chill. In the Tambora period, however, the presence of a sustained north wind fatally cooled the paddy water, upsetting the delicate balance of conditions for rice production. August temperatures in Yunnan for the three seasons from 1815 to 1817 were as much as 3°C below the seasonal average. This might not sound like much. On any given day, the difference between 15°C and 18°C (the low benchmark for optimum rice growth) is barely felt on the human skin. But seasonal averages are crucial indicators for temperature-sensitive crops. Every 1°C decrease in average summer temperature will reduce the rice growing season by up to three weeks, or the equivalent of a five-hundred-foot rise in elevation. At 3°C below average, almost no grain remains to be harvested at summer's end.

By the end of 1815, the meteorological impacts of the Tambora eruption were unfolding across Yunnan with dismal inexorability. After flooding volcanic rains in the summer, which destroyed both the spring and autumn harvests, food shortages gripped the province. The price of rice skyrocketed to 1,800 copper coins a bag, well beyond the reach of ordinary peasants. In Guanyin valley to the west, the villagers resorted to eating soil, an undigestible yellowish loam nicknamed "Guanyin noodles." Many died from painful swelling of the gut.

After a traumatic winter with high mortality, the people's hopes rose again in the late spring with the coming of the rains, this time in normal, moderate quantities. But the summer, of which these rains would ordinarily have been the sweet harbinger, never prospered. Instead, the bewildered and heartbroken Yunnanese endured unprecedented snows in July. In place of sunshine, the critical months of late summer and early autumn brought incessant rain, together with a meteorological phenomenon never before witnessed: great rolling icy fogs that lasted days on end. Even as late as the end of July, hopes persisted for a late harvest. But August brought a fresh onslaught of frosts and wintry gales, and the rice crop failed again, this time completely. The price of rice jumped once more into the thousands.

For the survivors of the first two years of famine, 1817 offered some initial relief from outright starvation. Parts of western Yunnan experienced snow for the first time in memory, but in the mountains of the more populous central region, the wheat and broad bean crops ripened promisingly. Hungry villagers rushed to dig up the beans and fill their empty baskets. But the rate of mortal starvation, which slowed for some months, accelerated again when, for the third year running, the vital summer months were cruelly cold. Snow fell again over Kunming and frosts covered the ground from June through August. The farmers of Yunnan—crop scientists all—had planted five different strains of rice, each calibrated to specific temperatures and elevations. But none of them was hardy enough for Tambora. By now, the disaster crippling Yunnan had the attention of the Qing court, thousands of miles to the east. In the autumn of 1817, the emperor's remote subjects on the southwest frontier faced their third comprehensive rice-crop failure in succession, and the greatest crisis in their history.

THE POETRY OF FAMINE

Qing China is best described as an autocracy with shallow meritocratic structures. Central to its organization were the imperial examinations, conducted nationwide, in which Chinese subjects regardless of class or background could compete for academic distinction and a government appointment, which was the surest path to upward social mobility.[10] The Confucian administrative ethos of popular welfare embodied in the celebrated granary system also permeated Chinese culture and education: a philosophy of governance in which the state ensured its ruling elite was thoroughly steeped. Confucian texts thus dominated the bureaucratic curriculum, in which students learned the importance of filial loyalty, social service, and self-sacrifice. Literacy also featured prominently and was tested in the form of poetic composition. The "mandarins" produced by these examinations were trained as scholar-bureaucrats, with a literary polish to their education comparable to the requirements of the elite European gentleman of the same period.

The imperial examination system, finally dismantled at the beginning of the twentieth century after seven hundred years, has been blamed for China's sclerosis in its "century of humiliation," for the inability of its ruling elite to respond effectively to the forces of modernity brought by aggressive Western states. An internal weakness of the system lay in its vanishingly small reward structure: only a handful of examinees in any given year were offered government postings, leaving the rest to private sensations of bitterness and failure that were potentially at dangerous odds with the precepts of community harmony they had worked so hard to imbibe. Having learned the necessity of social hierarchy in the Confucian system, unsuccessful examinees very often faced the reality of their own place in its lower ranks.

A Yunnanese man named Li Yuyang was one such disappointed Confucian student in 1815.[11] But his story was so close to being one of emblematic success for the empire. By birth he belonged to the Bai ethnic minority in Yunnan, a hill people brought under the wing of the Chinese state and encouraged over many generations to adopt the language, manners, and aspirations of their conquerors. During the active colonization of the southwest during the Ming dynasty, Li Yuyang's Bai forebears had been forced from their fertile lowland villages into the hills so that Chinese agriculturation of the land could take its remorseless course. But imperial assimilation policies over time assuaged the family's resentment. Sinicization was sufficiently advanced in the family of Li Yuyang for him to leave his home in the ancient capital of Taihe, near Dali in northwest Yunnan, for a prestigious Confucian academy in Kunming. It was a remarkable and ambitious career move for a young Bai. And, at first, he found success. In bustling Kunming, Li Yuyang studied under a well-known *mandarin* master, gained notice as a composer, and joined an exclusive group of poets under the master's tutelage. This group, of which he was the star, became known as the Kun Hua Five after the student neighborhood they inhabited.

But at some point, the aspirations of this rising would-be *mandarin* to complete his evolution from ethnic provincial subject to imperial Chinese ruling class began to falter. He failed, year after year, to achieve the necessary distinction in the imperial examinations. Genteel poverty was expected, even celebrated among the literati, as it was among the bohemians of faraway Paris and subsequent generations of the urban creative classes in the West. But at some point Li Yuyang's family went bankrupt, forcing him to leave his beloved academy and its neighborhood of intellectuals. He moved to the outskirts of Kunming, where he worked as an ordinary small-acre farmer in the rice fields alongside illiterate peasants. He had not long joined the ranks of the peasantry when Tambora's Frankenstein weather hit Yunnan, bringing chaos and death.

As the Tambora disaster began to unfold, Li Yuyang was thirty-two years old and in the process of constructing a new identity for himself as a poet of the people.[12] In September 1815, during the first blighted post-Tambora harvest, Li Yuyang looked back over the disastrous summer with its cold winds and incessant rain. By his account, the intensity of the volcanic downpours damaged houses and brought flash flooding to low-lying villages near Kunming. Here is his first Tambora poem, titled "A Sigh for Autumn Rain:"

The clouds like a dragon's breath on the mountains,
Winds howl, circling and swirling,
The Rain God shakes the stars, and the rain
Beats down on the world. An earthquake of rain.
Water spilling from the eaves deafens me.
People rush from falling houses in their thousands
And tens of thousands, for the work of the rain
Is worse than the work of thieves. Bricks crack. Walls fall.
In an instant, the house is gone. My child catches my coat
And cries out. I am running in the muddy road, then
Back to rescue my money and grains from the ruins.
What else to do? My loved ones must eat.
There are no words for the bitterness of
An empty September. The flood-drowned fields
harvest three grains for every ten of a good year.
And from these three grains? Meals and clothes till next September.

In Chinese folklore, the cloud dragon represents life-giving rain from the east, the crop-sweetening spring rains. But Li Yuyang conjures here an angry, unpredictable weather Dragon, roused to punish the people.

Environmental historians have linked the vulnerability of the flood-prone agricultural zones of Qing China to the policy of unconstrained logging and land clearance operated over several centuries. While the intensification of agriculture, including widespread double cropping, had to some degree decoupled crop yields from normal climatic variability, the manufactured agro-ecosystem was much more vulnerable, at the system level, to high-impact climate events such as Tambora. Bigger yields meant a bigger society, but it also meant that many more lives were now exposed to the dangers of a failed harvest.[13] Yunnan, ordinarily an exporter of grain to the rice-poor east, now faced a drastic food shortage of its own.

In a subsequent poem, Li Yuyang turns an accusing eye on local officials, who show no mercy even in the time of crisis, demanding the people pay their taxes as usual. Meanwhile, the legendary *Shang-yang*, bird of rain, still flies:

Grain tax! the policemen shout. Their whips
Slice the air, and the agony of the people
is neverending. Who can plug heaven with a stone,
or command the *Shang-yang* stop flying?
If only the sun would rise where it should,
And the dragon with his dark clouds disappear.
O our free hearts then! But when I ask if tomorrow
will be fine, the flower under my feet says nothing.

As the bad weather continues into the fateful year of 1816, Li Yuyang frets increasingly about the crop-destroying rains. He writes of feelings of helplessness as he watches his wife and children waste away from hunger. He takes to sitting up alone at midnight. Brooding over the painful images of the day, his hair begins turning white:

Rain falls unending, like tears of blood
from the sentimental man.
Houses sink and shudder
like fish in the rippling water
I see my older boy pulling at his mother's skirt.
The little one cries unheard. Money gone, and
Rice rare as pearls, we offer our blankets to save ourselves.
A single *dou* of grain, and nothing over to fix the house.
We have only a few acres, and these grow nothing.
My wife and children portion out their grains across
The wide year. At least the taxman stays away.
How could anyone fill his deep pockets?

Local sages, steeped in the "moral meteorology" of Confucian philosophy, were quick to blame the bad weather on some lapse in the conduct of the people, in the loyalty of sons or the chastity of daughters. Emperor Jiaqing, for his part, blamed the incompetence of the administrative class beneath him for compromising the goodwill of Heaven toward the state. In an imperial edict, Jiaqing explicitly deflected responsibility for the 1816 food crisis onto "provincial officials," who had they managed affairs diligently and in a completely public-spirited manner, all cooperating with each other, there would not have been a situation such as this. . . . The wheat harvest has already proved deficient. If the great fields are not sown in good time, there will be no supply of food for the humble folk.[14]

Spin doctoring aside, the very existence of such an announcement from the emperor shows a style of paternalistic noblesse oblige in the Chinese social contract unfamiliar to the recently restored monarchs of Europe. The obligation to express sympathy for the peasants in times of crisis held true along the entire chain of command—from the emperor to the local magistrate level where scholar-bureaucrats frequently published poems that memorialized the suffering of the people. Anthologies of Qing-era poetry contain entire sections titled "Famine and Calamities." Li Yuyang's poetry thus takes its place within an established paternalistic tradition in Confucian culture.

From Li Yuyang's reference to absent tax collectors, it seems that by late 1816 the provincial government of Yunnan had taken some measure of the humanitarian disaster and turned its attention from taxing the people to saving them. It's not difficult to see why. Li Yuyang's poems from the following year, 1817, turn increasingly from the desperate situation in his own home to the epochal human tragedy unfolding around him. The streets of Kunming present a spectacle of suffering and social breakdown to which no right-minded Confucian official could be immune. Desperate parents have begun bringing their children to market for sale:

> 300 copper coins for a bag of grain
> 300 copper coins for three days of life
> Where can the poor people find such money?
> They barter their sons and daughters on the streets.
> Still they know the price of a son
> Is not enough to pay for their hunger.
> And yet to watch him die is worse.
> Think of our son's body as food, as grain for one meal.

The little ones don't understand, how could they?
But the older boys keep close, weeping.
Stop crying and go with him. Selling is
a blessing, because to buy you he must feed you.
The cold wind blows in their faces,
The parents wipe their tears away.
But back home they cannot sleep
While the birds moan like old men in the night

Li Yuyang devotes another poem to an heroic minor official named Liu, who at this critical time took action to dismantle Kunming's new slave market for children. Liu hunted down the buyers and forced them to return their purchases to their families, including the daughter of a blind man who duly "prayed for his sight to be healed / if only to turn his grateful eyes upon Magistrate Liu."

By early 1817, the rising death toll and food panic in Yunnan had grown into a full-scale human emergency, forcing the government to open the granary doors and dispense its precious reserves for free. State-organized famine relief, essentially unknown in Europe until the twentieth century, stands as one of the greatest achievements of premodern Chinese civilization. All across the wide empire, for hundreds of years, Chinese officials perennially regulated the price and supply of its people's staple food by purchasing rice in the abundant autumn, storing it in state granaries, then selling its reserves in the winter and spring as supplies dwindled and prices rose.[15] Because of its isolation, Chinese rulers perceived Yunnan as a region particularly susceptible to food shortages; it was accordingly well-supplied with granary stock.[16] By the time of the Tambora emergency, reserve granaries had functioned in Yunnan for over a thousand years. The granaries of Yunnan, at least according to the officials who managed them, stored a one-month supply of food for every grown man in the province, the highest ratio in the empire.

But there may be reason to doubt these official figures. Increasingly in the early nineteenth century, disturbing reports reached the Peking court describing the dilapidated state of provincial granaries, which may itself have been the consequence of an imperial policy of

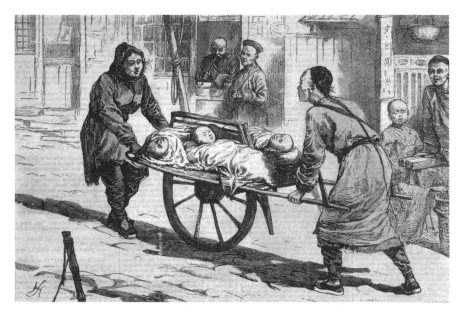

FIGURE 5.3. Europeans were scandalized by reports of starving children brought to market to be sold for bread during the Great Chinese Famine of the late 1870s—as dramatized in this newspaper illustration. As we have seen in the case of the Yunnan famine of 1816–18, however, it was a culturally accepted last resort for Chinese families facing starvation. (© The Granger Collection, New York).

deliberate neglect. The state management of the granary system proved so expensive that bureaucrats looked more and more to the grain markets as a means to rationalize food distribution. Why stuff every province with reserve grain when an efficient marketplace, in time of crisis, could transport it to the disaster zone according to the logic of supply and demand?

By 1815, the Qing state, ever fearful of the social instability wrought by food shortages, had come to favor a hybrid model of famine risk management that combined an integrated, commercialized food distribution network—that is, a grain market—with a long-established state-run granary system to guarantee food supply to its frontier peoples. This sophisticated model of grain distribution, evolved over centuries, proved highly successful under conditions of normal climatic variability. But when Tambora's 1815 eruption brought an unprecedented wave of extreme weather to the region—possibly the worst of the millennium in Asia—the whole elaborate system quickly cracked under the strain. That very year, 1817, an official edict to the Qing court lamented the state of the national granaries, pointing to years of neglect and bad management.[17] The timing could not have been worse for the suffering people of Yunnan.

In his poem about relief operations, would-be mandarin Li Yuyang swallows his pride and joins the starving crowds at the main gate of Kunming, where the granary managers set up tables and force the people into orderly queues for servings of a weak rice porridge. The vital work is poorly managed, and supplies are inadequate, but the emperor's charity is now the only recourse for a starving people:

> You open the Li Gate, and the hungry millions moan
> At the smell of gruel. You give a bowl to the grown man,
> half to the child. But don't you see the strong men push forward,
> while the old stumble? We wait until noon,
> Bellies hollow like thunder. But your porridge
> Is like water. I will come again tomorrow,
> if I am not already dead. I will beg again
> For porridge, but quietly, so not to anger you.

The "porridge" was poor stuff indeed, consisting of barley flour and broken rice seeds mixed with buckwheat or vegetables—a deliberately wretched potage so that only the truly famished would line up to consume it.

Given that the standard granary reserves in Yunnan were sufficient to feed a maximum of 15% of the population at any time and that the years preceding the Tambora emergency were drought years, it is not surprising that the government's means to stem the famine were soon exhausted. In "Bitter Famine," Li Yuyang describes the food crisis at its worst in the autumn of 1817, as the people of Yunnan descend into a living hell, their prosperous communities transformed into a Dantean circle of starvation and death, but with no innocents spared:

Outside, the starved corpses pile high,
While in her room the young mother
Waits upon her child's death. Unbearable
Sorrow. My love, you cry to me to feed you—
But no one sees my tears. Who can I tell which aches
More? My heart or my body wasting away?
She takes her baby out to the deep river.
Clear and cool, welcome water . . .
She will care for that child in the life to come.

Confucian values focus on the sacred debts of children to their parents who have dedicated their lives to their offspring's welfare and protection. The infanticide that concludes this poem thus makes for a wrenching irony. The young mother fulfills her Confucian duty only by drowning her child and herself.

With the food situation at its gravest, Li Yuyang wrote another bitter poem about family loyalty, this time of a poor man whose filial virtues go unrewarded. He sacrifices his own family to feed his mother, according to his Confucian bond, but then dies himself anyway, leaving his mother alone and desperate:

Around the neighborhood, you can hear her crying,
That old widow, cold and hungry, and in rags.
She will tell you the famous story of her son:
His tireless hands, in the fields from dawn to dusk,
Could only feed two mouths. He took care of
His mother, hence his fame. But Death cared less
And took him away. The sweet bonds of their love
Untied, the grey-haired widow is all alone.
"My time is short, and I dream for our reunion,
But life teases me awake. Why do I still have breath
When I have no food? Take me, for his soul's peace! . . ."
But only the birds listen. They take flight into the darkness.
Lucky birds, however distant, fly home. Not her.

The early months of 1818 brought no relief from Tambora's suffo-cating grip on the grain-growing seasons of Yunnan. A recent mod-eling study on the impact of Tambora's eruption on the Chinese cli-mate found that the coldest temperature anomalies occurred not in 1816 but 1817–18, which "may explain long lasting impacts like the three years famine in the province of Yunnan."[18] Indeed, several studies on Tambora's influence on Himalayan weather to the northwest point to the eruption's great reach, in both space and time, extending cold tem-peratures into the 1820s.[19] In the mountain city of Kunming, Li Yuyang writes of a heavy snowstorm in January 1818, complete with lightning, thunder, and "purple rain" that blasted the winter crops of broad beans and wheat. This is the last of his famine poems. Now well into their third successive year of dearth, the suffering of the Yunnanese in early 1818 may well have passed beyond description for Li Yuyang.

Mercifully for the survivors, this was to be the last of the Tamboran crop failures. By the summer of 1818, the volcanic dust had at last cleared from the stratosphere, and the sun and balmy southwest rains returned as normal to the land "south of the cloud." A bumper crop that autumn brought an end to Yunnan's long despair. As for Li Yuyang, he survived the great famine in body, but there are signs of a permanent trauma of spirit. His brief official biography tells of an increasingly reclusive man "who never left the inner door of the house, and died at home."[20] Sitting up at midnight during the dark days of the Tambora disaster, Li Yuyang felt the white hairs sprouting from his head. Prematurely aged by the suffering to which he had borne such eloquent witness, he died in 1826 of pulmonary failure, aged forty-two.

THE OPIUM CONNECTION

Thousands of miles from Li Yuyang's family in crisis and the unfolding disaster in Yunnan, Fanny Godwin's state of mind was deteriorating through the dreary summer months of 1816. While her sisters, Mary Shelley and Claire Clairmont, seized their roles abroad in an emerging literary and cultural revolution, she stayed at home in London to bear the complaints of her father and a hostile stepmother. This oppressive isolation, together with her unrequited feelings for the irresistible Percy Shelley, brought sadness and anxiety in waves upon her. The Geneva party had returned to England, without Byron, in early September 1816; but Mary and Percy avoided the unhappy Godwin house in London, instead staying in Bath. Fanny sought them out and, on October 8, met alone with Percy. Whatever passed between them, his cool, ambiguous behavior was the final straw for the abandoned Godwin sister. In a sad, regret-filled poetic fragment, Shelley later recalled how "Her voice did quiver as we parted,/Yet knew I not that heart was broken." Fanny immediately left Bath, traveling on to Wales. The following day, in a Swansea hotel, she scribbled a note blaming herself and asking her loved ones to forget her. The chambermaid found her dead the next morning from an overdose of opium.

The fact that a respectable and inexperienced young woman in Britain chose suicide in the form of a half bottle of laudanum demonstrates both opium's easy availability in the immediate post-Waterloo years and the fact that, though valued for its medicinal properties since ancient times, the poppy's dangers were not yet widely appreciated or regulated. In 1816, most English opium was imported from the Near East along the trading routes of the Mediterranean. But since British deregulation of the Indo-Chinese trade in 1813, the global market for opium had expanded rapidly, while the center of production and consumption shifted to the Far East. By 1827, Britain's success in penetrating the Chinese market for opium had reversed the flow of silver between the trading partners, which had so long been to the advantage of the Chinese.

From that point, the long-powerful Chinese empire suffered a series of crushing setbacks through the nineteenth century and beyond. It lost its leading role in world trade to Britain, certified by the ruinous terms of surrender that concluded the Opium Wars in 1842 and 1857. Consequently, per capita income for its citizens actually declined through the nineteenth century while the Euro-Atlantic zone raced ahead in economic growth and technological advancement.[21] For the Chinese

Communist Party rulers of the 1950s, looking back over the ruins of the "century of humiliation" that followed China's first defeat by Britain, opium was to blame for the civil strife, famines, and military defeats that had ruined China's once great empire and thrown the country into economic and social chaos. This anti-Western narrative—a pillar of Communist Party historiography—focused on the evils of *imported* opium, a market Great Britain had unscrupulously created and kept open with military force.

Our Tambora story, however, takes us further back in time in the history of Chinese opium to the site of what would become the thriving center of *domestic* opium production in the Qing empire: the southwest provinces at China's frontier. The Qing court had long been concerned with the importation of Indian opium by the British and sought to control the trade along its southern ports. Beginning in 1820, however, only two years after the end of the Tambora-driven famine, Chinese rulers in Peking were startled to receive reports from faraway Yunnan of a sudden explosion in opium production there. A poppy anticultivation program was instituted for Yunnan that very year, the first in a series of ever more desperate government measures to curb the southwestern drug industry. But to no avail. Opium in ever greater quantities continued to flow south along the Red River into Vietnam transported by enormous caravans, and from there by sea to Hong Kong and Canton, or eastward overland through China. Nothing could stem the tide, and by 1840—during the first Opium War with Britain—Yunnan was the acknowledged heart of domestic opium production in China. Not all of Yunnan's opium left the province, of course. By this time, more than half of Yunnan's garrisoned soldiers were reckoned to be drug users, including the officer corps.[22]

What caused Yunnan's sudden transformation, in less than two decades, from a grain-producing province well integrated with the empire's agricultural system to a rogue narco-state in thrall to the international drug trade? With Tambora's specific dates in mind, what follows is a probable scenario.

By the time of the Tambora emergency, the commercialization of agriculture in southwest China had evolved to the point where

FIGURE 5.4. This British illustration dating from China's humiliating defeat in the first Opium War (1839–42) puts a benevolent face on opium addiction. The mood in this Chinese "opium den" seems recreational, even festive. Not visible in the image is an acknowledgment of Britain's vital trade interest in expanding its market of Chinese drug users or the devastating long-term effects of mass opium consumption on Chinese society. (Thomas Allom, *China in a Series of Views Displaying the Scenery, Architecture, and Social Habits of That Ancient Empire* [London, 1843–47], 3:54).

self-sufficiency was not the dominant working rationale of the common Yunnanese farmer. Rather, he found himself forced into the marketplace to raise money for taxes and buy grain in the off-season. In this light, the state bureaucrats who habitually railed to the court against the "stupidity" of the peasants for selling their excess harvest rather than storing it as a wedge against crop failure appear disingenuous indeed. For the low-acreage farmer subject to this commercial market, and in the teeth of a famine, opium must have represented an irresistible temptation: the poppy was worth twice as much per acre of yield than the average grain crop and would grow in inhospitable conditions on marginal soil. Sown in the fall, the opium flower grew to maturity in March and could be harvested for its sap in summer. It could thus to some degree be grown in conjunction with, or as supplement to, conventional food crops. At a critical point in the late 1810s, after years of the worst famine in their experience, the desperate peasant farmers of Yunnan came to the collective realization that opium was as good as money and more reliable than food.

Whatever the advisability of large-scale conversion of land to opium production from the empire's point of view, for the individual freehold farmer of Yunnan, food security was best served by significant investment in opium as a security against grain shortfalls and the recurrence of famine. Just as important, growth in poppy production served the interests of unsalaried local officials themselves, who were under pressure to meet tax revenue both to pay their own wages and to remit quotas to the court. The empire's long-successful system of provincial revenue extraction failed to adapt to the drastic climate change episode of 1815–18 after which the lure of opium as a cash crop was overwhelming. Once the opium land conversion had occurred, officials had no incentive to enforce anticultivation measures when they could tax the lucrative crop instead. In 1820s Yunnan, the foxes took guard of the opium henhouse.

In short, faced with multiyear food shortages in the Tambora years of 1815–18, Yunnan's farmers found they could neither grow rice nor buy it when they most needed it. Circumstantial evidence suggests that they subsequently settled on an opium solution to their chronic food security problem, hence the explosion of poppy farming in Yunnan from the late 1810s. First, the mountain fields of bean and wheat were converted en masse to opium production. Thereafter, poppy growers made their way brazenly to the central valleys, to colonize the choice arable land of the province. Fast forward a century later, and Yunnan was growing almost nothing but opium, importing most of its rice from Southeast Asia. At this time, ethnic hill tribes from Yunnan, such as the Hmong, began to drift southward into the Mekong delta, to the mountains of modern-day Burma, Thailand, and Laos. With generations of experience in opium farming behind them, they brought with them the seeds and technologies to establish a new global capital of opium production in these remote highlands of the "golden triangle."

Thus the Tambora period marks not only the beginning of a complete transformation of Yunnan's agricultural economy from staple grains to an opium cash crop but also the first emergence of the modern international illicit drug trade. That this evolution began in the aftermath of the Tambora emergency shows the sinuous correlation that can exist between high-impact climate change events—such as a three-year famine—and social disruption on global scales and centennial time frames.

As the secretary-general of the National Anti-Opium Association of China reflected in 1935, in the midst of China's long, tumultuous civil wars following the collapse of the empire in 1911, "the weakening of the race and the rapid increase of social evils can in the last analysis be traced back to their source in opium."[23] The early twentieth century was a time when China held the dubious honor of exporting over 80% of the world's narcotics. In the same period, in the onetime Confucian stronghold and Qing-era boom state of Yunnan, 90% of adult males were drug users, half of them addicts. A Western observer gives a graphic account of the human tragedy of opium in early twentieth-century China at the village level:

> The roofs of the houses are dilapidated and full of holes. . . . No one is selling vegetables in the road, and the one or two shops which the village possessed are closed. In the shadow of the houses a few men and women are lying or squatting—apparently in a stupor. Their faces are drawn and leathery, their eyes glazed and dull. . . . Even some of the babies the women carry in their arms have the same parched skins and wan, haggard faces. And the cause of all this is opium.[24]

This description of an opium-afflicted community reads like a "Seven Sorrows" poem in the spirit of Li Yuyang. This long-suffering, long-forgotten writer—whose poems appear here for the first time in English—fulfilled his destiny as a Confucian poet of the people in memorializing the Great Yunnàn Famine of 1815–18. But he spent the remainder of his life in scholarly seclusion, as if in bitter meditation on the disturbing changes afoot in Yunnan and the national humiliation in store for his beloved China.

CHAPTER SIX

THE POLAR GARDEN

From the flooded mountain pastures of Yunnan, we must now travel thousands of miles aboard Tambora's sulfate plume to the melting ice cap of the polar north. As we have seen, the Arctic has been, since the 1960s, a key repository of scientific evidence for reconstructing Tambora's eruption and climate impacts. But in the years immediately following the 1815 eruption in Britain, "scientific" interest in the Arctic was inseparable from the twin political agendas of government and the Royal Navy: namely wealth and glory. How Tambora conspired to both excite and thwart the Arctic dreams of a generation of British bureaucrats and explorers makes for what is, in many ways, the strangest of all Tambora's strange tales.

A musty memo from a Royal Society council meeting in November 1817 would seem unlikely to resurface, two centuries later, as a text of quasi-biblical importance on the blog pages of the climate change denial community. But such is our own topsy-turvy world of politics, science, and Internet-driven opinion in the century of climate change. The Royal Society minute in question is attributed to none other than Sir Joseph Banks, aging lion of the British scientific establishment. In it he refers, in excited terms, to newspaper reports of a rapidly melting Arctic ice cap.

The audience for the memo was the First Lord of the British Admiralty, for whom Banks painted a seductive picture of an open polar sea through which the navy's ships might sail in quest of scientific discovery, Asian trade routes, and national glory:

> A considerable change of climate inexplicable at present to us must have taken place in the Circumpolar Regions, by which the severity of the cold that has for centuries past enclosed the seas in the high northern latitudes in an impenetrable barrier of ice has been, during the last two years, greatly abated. This affords ample proof that new sources of warmth have been opened, and gives us leave to hope that the Arctic Seas may at this time be more accessible than they have been for centuries past, and that discoveries may now be made in them, not only interesting to the advancement of science, but also to the future intercourse of mankind and the commerce of distant nations.[1]

In fact, these words were probably not composed by Joseph Banks at all but written *for* him by a senior bureaucrat at the Admiralty—the Machiavellian Second Secretary, John Barrow—for reasons that will become soberingly clear. Authorship questions notwithstanding, prominent climate denial bloggers have trumpeted Banks's description of Arctic warming in 1817 as historical proof that "the ebb and flow of Arctic ice extent and mass is nothing new" and, even more definitively, that "climate change is not a new phenomena [*sic*]."[2]

My main purpose in this chapter is to tell the story of Arctic environmental change in the aftermath of Tambora's eruption and its remarkable formative influence on the history of British polar exploration. But a bonus of the research presented here will be the opportunity to lay to rest—among the legion of moldy myths with which the Arctic north has for centuries been encrusted—the false notion that the ice-free polar seas of 1817 were a product of natural variability of climate and that, accordingly, we should think nothing of the catastrophic ice declines of the early twenty-first century.

What caused the breakup of the polar ice cap in 1817 that to Sir Joseph Banks was so "inexplicable"? What were the strange "new sources of warmth" that held out the tantalizing promise of an open polar sea? As the recent scientific literature on volcanism and climate clearly describes, Tambora's massive eruption in 1815 precipitated—through an extended physical chain of dynamic events involving earth, sea, and sky—a freak, drastic, but *temporary* diminution of Arctic sea ice well outside the bounds of normal variability. Fossil-fuel emissions of our industrial age have the same warming impact on the Arctic as volcanic sulfate aerosols (albeit by different mechanisms), but the influence of volcanic dust possesses the decided advantage—to the Arctic and to humanity—of disappearing after a few years. Twenty-first-century anthropogenic warming, by contrast, has no foreseeable time horizon and has set us on an inexorable course, like the heedless explorers of yore, toward a once unimaginable ice-free polar sea.

GLOBAL WARMING, NINETEENTH-CENTURY STYLE

In the wake of the bloody, manic conclusion to *Hamlet*'s revenge plot, the fallen prince's friend Horatio, surveying the Danish throne room littered with corpses, promises to "speak to the yet unknowing world / How these things came about." It's a difficult task given the confusion of what has just happened. Horatio alludes to "unnatural acts" and "accidental judgments," to "casual slaughters" and "purposes mistook."[3] In short, he emphasizes the contingency of events that have led to the royal massacre. The Tambora period, as I am chronicling that global tragedy in this book, is rife with such highly contingent events. The sudden outbreak of epidemic cholera in Bengal, for example, and the boom in opium production in the "golden triangle" region of southwest China post-1815 depended on a multitude of causes converging over time with an initial, triggering climate change event.

No episode in these calamitous years better conforms to Horatio's description, however, than the British Admiralty's decision to embark on a doomed quest for the northwest passage in 1818. Moreover, in Admiralty secretary John Barrow, we find an historical actor whose "accidental judgments" and "purposes mistook"—not to mention

FIGURE 6.1. Portrait of Sir John Barrow from about 1810, early in his forty-year tenure as a highly influential Second Secretary to the Admiralty. (© National Portrait Gallery, London.)

"unnatural acts" of journalistic propaganda—gave rise to a noble but ultimately tragic and fruitless Arctic adventure played out on the nineteenth-century stage.

Barrow, the most powerful bureaucrat at the Admiralty in the post-Napoleonic period, used the pages of the widely read *Quarterly Review* to advance his policy agenda for Britain's navy. Mary Shelley was staying at the Hampstead cottage of her friend the radical journalist Leigh Hunt when, among tea tables cluttered with the latest books and newspapers, she took up the February 1817 issue of the *Quarterly Review* to read a breathless review of the latest installment of Byron's poem *Childe Harold's Pilgrimage*, which featured scenes from their recent Genevan summer. Sitting right next to the review of Byron was matter of even more relevance to the summer of 1816 and to her writing of *Frankenstein*: a review by John Barrow chronicling the heroic history of Britain's quest for a northwest passage to Asia across the polar seas.

In the *Quarterly* article, Barrow floated his latest trial balloon from the backrooms of the Admiralty: that a select band of naval officers, legions of whom had languished in port since the defeat of Napoleon, should now be put to sea to renew the glorious search for a navigable passage to China across the uncharted top of the world. Despite centuries of failure in this quest, Barrow assured his readers that such an expedition would now be "of no difficult execution." Sailing westward across the north of Canada to the Pacific could amount to no more than "the business of *three months* out and home."[4]

Then came the jaw-dropping coincidence with which this history truly begins. Only months after Mary Shelley read Barrow's article in February 1817 calling for renewed exploration of a northwest passage, reports began reaching the Admiralty of a remarkable diminution of Arctic sea ice. The authority of these reports rested with the veteran whaler William Scoresby, who in 1815 had published the first scientific treatise on polar ice. The 1817 summer whaling season off the east coast of Greenland had been a ruinous disappointment. A frustrated Scoresby identified the cause in his journal: No ice!—

The fishery of the present season has been the most singular, partial, unsuccessful of any occasion witnessed of many years. . . . The ostensible reason of the scarcity of whales & their pecular [*sic*] habits, is the singular state of the ice which lies at a distance from the land greater than was ever known by any fisherman now prosecuting the business. . . . So thin is the ice dispersed through the country, that it is creditably asserted that a brig from the Elbe has penetrated without hindrance to the West land [the east coast of Greenland] and coasted along the shore to a vast distance & returned again eastward without difficulty, but without finding any whales![5]

FIGURE 6.2. The air of authority evident in William Scoresby's portrait, combined with his extensive knowledge of the polar regions, should have made him an obvious choice to lead the British Arctic expeditions of 1818. But Barrow snubbed him because he was a commercial whaler and not a navy man. In his absence, the Ross and Buchan expeditions floundered. (William Scoresby, *Account of the Arctic Regions* [Edinburgh, 1820]; © Bridgeman Art Library.)

On his return to England in August, Captain Scoresby, eager to justify himself in the eyes of his disappointed investors, published a short account of the circumstances of the poor whaling season in a Liverpool newspaper.

As we have seen, news of Arctic sea-ice loss reached the vigilant eye of Sir Joseph Banks, who had built his extraordinary career in the public promotion of British science on looking always to the next frontier. If Scoresby's reports were true, an ice-free Greenland sea was certainly a major development in the world of science, one the British nation should be quick to exploit. Banks dashed off a letter to Scoresby—whom he knew already as one of his many hundreds of correspondents on scientific subjects—asking for more "particulars." Even in this initial letter, Banks showed himself eager to theorize a synoptic connection between the melting of the polar ice cap and "the frosty springs and chilly summers we have been subject to" in 1816 and 1817.[6] Scoresby's tale dovetailed tantalizingly in Banks's mind with numerous reports of icebergs seen floating unnaturally far south in the Atlantic. A mile-long iceberg had been sighted off the Grand Banks while ghostly convoys of ice drifted past the coasts of Ireland and New York, and even, it was said, into the tropical Bahamas.[7] The miserable "Year without a Summer," so fresh in the memory, might be explained (Banks conjectured) by massive chunks of polar ice now drifting southward, cooling air temperatures along the way. Of greater national importance, of course, were the implications of an ice-free Arctic for a northwest passage to Asia.

Scoresby's reply to Banks's excited speculations was satisfyingly direct: "I found about 2000 square leagues [61,000 km²] of the surface of the Greenland sea, between the parallels of 74° and 80° north, perfectly void of ice, which is usually covered with it. . . . Had I been so fortunate as to have had the command of an expedition for discovery, instead of fishing, I have little doubt but that the mystery attached to the existence of a north west passage might have been resolved."[8] If Banks was thrilled at this report, what were the feelings of John Barrow, who had so recently staked his public reputation on the feasibility of a northwest passage? It must have seemed like manna from heaven. The two powerful bureaucrats soon joined forces to transform Barrow's wishful projections in his February review article into a fully funded reality. And in that moment, a famous Royal Society memo—holy writ to the modern climate denialist—was born.

In addition to Scoresby's account of an ice-free north, rumors filtering in from foreign sources were perfectly designed to light an Arctic fire under the Admiralty. The German navigator Otto von Kotzebue, in the employ of the Russian government, had been commissioned to explore the possibilities for a north*east* passage across the Arctic's Russian coast. In the mostly disastrous summer of 1816—the summer New Englanders called "Eighteen-Hundred-and-Froze-to-Death"—Kotzebue and his Russian crew, by contrast, enjoyed "delightful weather" and a clear passage through the Bering Strait, north of Alaska. As he records in his subsequent published account, on August 2, 1816, Kotzebue "sent a sailor to the mast-head [where] he announced that there was still nothing but open sea to the east . . . at which our joy was indescribable." Kotzebue clearly thought himself on the brink of a major geographical achievement, one that would place him in the ranks of Cortez and Cook: he "cherished the hope of discovering a passage into the Frozen Ocean, more particularly as the strait appeared to run without impediment to the horizon."[9] He was foiled, in the end, only by the shallowness of the local waters ahead that made further northeastward navigation impossible.

News of Kotzebue's near-success sent a shiver through the Admiralty. It would not do for Britain's navy, the most powerful in the world, to concede the prestige of Arctic discovery to a crew of Russian Johnny-come-latelies captained by a German mercenary. Would Britain, who had poured men and treasure into northwest passage exploration since the days of Queen Elizabeth, allow herself to be pipped at the post in the race to the Arctic? Clearly Lord Melville, the First Lord of the Admiralty, thought it must not be. He promptly gave his blessing to the Banks-Barrow plan to outfit two expeditions for the Arctic, the huge expense to be justified for "the national advantages which they involve but also for the marked attention they [call] forth."[10] In other words, profits and prestige. After all, the Royal Navy, with Nelson long dead and the French defeated, stood in desperate need of a new mission and fresh heroes to maintain its preeminence in the public eye.

Armed with the First Lord's stamp of approval, Barrow rushed back into print to announce the new naval missions to the Arctic. The issue of the *Quarterly Review* that appeared in February 1818 sold twelve thousand copies on its first day of publication, a record amount. Barrow's article—its major selling point—displays all the giddy self-confidence one might expect from a man who, having recently floated the notion of a passage through the Arctic, hears that vast territories of ice have promptly vanished, as if a slave to his own genius and will. Now, in the blink of an eye, a flotilla of ships crowded with glory-hungry men stood at his command.

It is with Faustian bravura, then, that Barrow's article presents a sweeping scientific and historical rationale for the government-funded polar enterprise. He proclaims to the British public the news of the permanent melting of the polar ice cap and the commencement of a new Golden Age of planetary warming:

> Among the changes and vicissitudes to which the physical constitution of our globe is perpetually subject, one of the most extraordinary, and from which the most interesting and important results may be anticipated, appears to have taken place in the course of the last two or three years, and is still in operation. . . . The event to which we have alluded is the disappearance of the whole, or greater part of th[e] vast barrier of ice.[11]

Barrow proceeds on an ambitious survey of the history of climate of the last millennium, beginning with what is called today the Medieval Warm Period, when "vineyards were very common in England" and colonists from Denmark and Norway settled the south coast of Greenland. When the Little Ice Age subsequently descended on Europe and North America, a new regime of cooler temperatures enlarged the empire of northern ice, cutting off the Nordic settlements and closing the northwest passage. Since 1815, Barrow continues, a further drastic drop in temperatures has killed crops across the hemisphere and fueled the Alpine glaciers (he is well-informed of developments to be described in the next chapter). But this last "deterioration of climate," Barrow argues, is cause for celebration because it represents the last gasp of the four-hundred-year cooling regime over Europe.

How did Barrow reach this original conclusion? "It can scarcely be doubted," he argues, that Europe's recent string of cold summers has been owed to the presence of Arctic icebergs drifting southward en masse in the Atlantic. Once these ice floes have melted in the southern latitudes, as they must, and the Arctic is ice free, his readers might look forward to "once again enjoying the genial warmth of the western breeze, and those soft and gentle zephyrs, which, in our time, have existed only in the imagination of the poet."[12] An open Arctic will transform foggy England into a sunny Arcadia.

Barrow should not bear full responsibility for his trumped-up utopian views of climate change in 1817. Percy Shelley himself had dared to imagine something very similar only a few years prior in his revolutionary poem *Queen Mab*. Shelley's Fairy Queen looks forward to an era of global warming in which ice caps are "unloosed," and changing wind and ocean circulation usher in a new climatic regime "full of bliss" for humankind:

> Those wastes of frozen billows that were hurled
> By everlasting snow-storms round the poles,
> Where matter dared not vegetate or live,
> But ceaseless frost round the vast solitude
> Bound its broad zone of stillness, are unloosed;
> And fragrant zephyrs there from spicy isles
> Ruffle the placid ocean-deep, that rolls
> Its broad, bright surges to the sloping sand,
> Whose roar is wakened into echoings sweet
> To murmur through the heaven-breathing groves
> And melodize with man's blessed nature there.[13]

Barrow's conservative *Quarterly Review* loathed Shelley's radical poetry; the editors called it "satanic." But by 1817, the educated classes were steeped in the new earth science of Buffon and Cuvier, who promoted the idea that Earth's long history included episodes of radical environmental change. Pastoral images of perpetual summer and polar gardens proved irresistible to writers of all political persuasions. An unfrozen north meant prosperity and freedom, perhaps even a revolution in consciousness. Few except those with firsthand knowledge of the brutal polar seas—plus natural skeptics like Mary Shelley—stood outside the Arctic climate change consensus of the late 1810s.

Barrow spends little time in his remarkable essay exploring what has caused the "revolution" in the Earth's climate since 1815. He borrows Benjamin Franklin's theory of electrical atmosphere to suggest the aurora borealis may be responsible for melting the ice, but he soon throws up his hands. It is "enough," he concludes, "to consider it as the result of one of those prospective contrivances, which are appointed to correct the anomalies, and adjust the perturbations of the universe."[14] Beneath the cocksure rhetoric lies a set of hollow presumptions. For example, Barrow offers no evidence for his assumption that the prior four centuries of cooler temperatures—the Little Ice Age—constituted a climatic anomaly, nor for asserting that the current trend toward diminished sea ice, based on a slim few years' sample, might represent a permanent benign change. In place of reason, as Mary Shelley well saw, Barrow offered only gauzy romance—a grand illusion papered over with shreds of truth.

BERNARD O'REILLY: THE FORGOTTEN MAN

The accounts of Scoresby and Kotzebue of open Arctic waters in 1816–17, gift-wrapped by John Barrow as a glorious prize to be claimed for the nation, was sufficient to launch a flurry of polar discovery teams in the years that followed. Naval officers John Ross, David Buchan, William Edward Parry, and John Franklin had all embarked upon highly publicized expeditions by the decade's end. Their extreme experiences en route and fluctuating reputations on their return set the terms of heroism, ignominy, suffering, and inconclusive defeat that were to drive the British Arctic narrative for the next half century. But in all the hoo-ha, one vital account of the highly unusual state of the Arctic in the aftermath of Tambora has been almost entirely forgotten (thanks to John Barrow). This polar journal, from the summer of 1817, is especially important since it provides eyewitness testimony of open waters in the aftermath of Tambora's eruption to the *west* of Greenland, in Baffin Bay, whence the iconic expeditions of Ross, Parry, and Franklin would later begin their quests for the northwest passage.

It is too often a crutch for historians to describe a little-documented figure from the past as "obscure." Obscure to whom? But there can be few other words to describe Bernard O'Reilly, a young Irish naturalist of unknown lineage who bobs up into public view in the 1810s. Traces of O'Reilly's stillborn scientific career may be found in the archives of the Dublin Royal Society. But his major bid for fame takes the form of a book, a single volume published in London in 1818, boldly titled *Greenland, the Adjacent Seas, and the North-West Passage to the Pacific Ocean.* From the opening page, O'Reilly announces his ambition for the book to redress the scandalous "want of scientific information on the northern climates."[15]

As we have seen with William Scoresby's 1817 summer voyage—plying the waters east of Greenland on a fruitless search for whales spooked by lack of ice—Bernard O'Reilly, hitching a ride on a British whaler, had chosen a strange, historic year to explore the Arctic. He found the western waters off Greenland—Davis Strait, Baffin Bay, and Lancaster Sound—equally void of ice. The polar ice cap was clearly visible to the north beyond 78°, but in the direction of the putative passage to the Pacific, with "all the broken field ice having drifted down to the southward . . . the sea remain[ed] as clear as the Atlantic, blue, and agitated by a considerable swell from the north-west!"

In one tantalizing passage from O'Reilly's account, he records the experience of a veteran captain that summer who, having ventured farther northwest than any whaler before him, observed in amazement from the masthead an unprecedented invitation to "proceed as far north as he pleased" across Melville Bay, on a "heavy open sea" with "no obstruction."[16] History turns on small moments, with no band playing.

Just then, the captain remembered his sacred whaler's oath to pursue no object but whales and so turned his ship about.

Even so, O'Reilly was able to confirm Scoresby's account: that "owing to some convulsions of nature, the sea was more open and more free from compact ice than in any former voyage they ever made … that, for the first time for 400 years, vessels penetrated to the west coast of Greenland, and that they apprehended no obstacle to their even reaching the pole."[17] Bernard O'Reilly's book should have made him famous. But the new field of Arctic literature had quickly gotten crowded, and he found himself elbowed sharply from the scene by the powerful figure of John Barrow, who himself published a book on the Arctic in 1818.

Barrow's achievements as an Admiralty bureaucrat over the previous twelve months had been impressive to say the least. In that time, he had turned a book review and a few whalers' reports into two fully manned expeditions to the Arctic Ocean and a spectacular publicity coup for the Royal Navy. Swapping on his naturalist's hat, he had also publicly announced a new benign era of global warming. As the Arctic expeditions, under the command of Captains Ross and Buchan, made their preparations in Portsmouth at the beginning of 1818, Barrow had only loose ends to tie up.

Bernard O'Reilly was one such loose end. In a *Quarterly Review* article of April 1818, Barrow eviscerates O'Reilly's *Greenland* as among "the most barefaced attempts at imposition which has occurred to us in the whole course of our literary labours," a book full of "nonsense," "falsehood," and "glaring folly," and motivated throughout by a "mischievous tendency."[18] At first blush, Barrow's vituperation of O'Reilly seems perplexing. After all, had not O'Reilly done great service to the naval cause in publishing his eyewitness account of an open polar sea in precisely that region to which Captain Ross was now preparing to sail in search of a northwest passage?

Fatefully for him, however, O'Reilly openly rejected the notion of a northwest passage to be found anywhere but along the Canadian north coast (where no expedition had been sent) and flatly rejected Barrow's theory of polar warming and an open Arctic sea. Worse still,

O'Reilly dared to mock Barrow himself for believing in such chimeras. He called the new Admiralty expeditions a "utopian paper-built plan," destined to "futility":

> Sailing to the north pole has been long a very favourite subject for closet lucubration; and as long as a man, in such circumstances, chooses to amuse himself harmlessly, or entertain his friends with his effusions through the medium of a magazine, such pursuits are altogether allowable; but where such visionary schemes are in contemplation as would mislead the public mind, in the same manner as the writer misleads himself, not pausing over facts, and maturely weighing their consequences, the prudent will be careful how they admit his opinions, however plausibly dressed up.[19]

O'Reilly (like Mary Shelley, as we shall see) reveals himself a polar skeptic, a Barrow critic, and hence an enemy of the Royal Navy. And because the impudent Irishman brought to his skepticism the authority of eyewitness, Barrow—who had never been to the Arctic—calls him a fraud and a plagiarist, a peddler of "absurdities, too mad for reason, too foolish for mirth."[20] The final straw? O'Reilly had the gall to publish his book as a lavish, expensive quarto targeted to Britain's elite circles. From Barrow's point of view, the claims of uppity amateur naturalists over Arctic science must be denied as emphatically as those of professional whalers—Scoresby, too, found himself unceremoniously excluded from the new polar mission. Only Britannia, in her full naval glory, should rule the Arctic waves.

Barrow's scathing review did its work well. O'Reilly's *Greenland* disappeared from sight, crushed beneath the wheels of Barrow's Arctic juggernaut. In subsequent years, this unfortunate Irishman traveled the south seas as a ship's surgeon, compiling scientific observations on Java, Australia, and India for future volumes that never appeared. Bernard O'Reilly was found dead in a rented room in London in 1827, probably by his own hand, his desk cluttered with desperate letters to would-be patrons. In one, he takes credit for Britain's renewed quest for the northwest passage, a blatant untruth he nevertheless had good reason to believe might have been true ... but for his nemesis John Barrow.[21]

CAPTAIN SCORESBY'S MARINE DIVER

So how did tropical Tambora—distant and faceless—launch her thousand ships into the Arctic? More pointedly, how could a period of drastic global cooling, precipitated by Tambora's eruption, be consistent with warming of the polar seas and the release of thousands of square miles of Arctic ice into the shipping lanes of the Atlantic? William Scoresby, unwittingly, provides a key piece to the puzzle in his whaling journal of 1816.

Scoresby cut a legendary figure among the whalers of Yorkshire for his superb seamanship and physical charisma: he was capable, it was said, of staring down polar bears. In addition to these Ahab-like acquirements, he was a naturalist of rare gifts. In the whaling off-season, he studied with the renowned natural historian Robert Jameson at the University of Edinburgh and corresponded, as we have seen, with Sir Joseph Banks. His compilation of a lifetime's study of the polar region, published in 1820 as *An Account of the Arctic Regions*, remains a classic foundational text of Arctic science. Charles Darwin, a fellow student of Robert Jameson, kept a well-thumbed copy by his bedside onboard the *Beagle*.

The main theme of Scoresby's correspondence with Banks prior to 1817 is the design of an instrument for measuring seawater temperature at depth. Banks commissioned a slim bucket made of glass and wood for Scoresby to collect samples, but the whaler found that, at three hundred fathoms, the wood swelled and broke the glass. Scoresby then himself designed an upgraded model of the water sampler—an elegant, self-closing, octagonal cylinder made of brass, with a single window. The contraption weighed twenty-three pounds and sank by itself without the need of ballast. Scoresby called it his "Marine Diver." It was Scoresby's habit to take advantage of any lull in the hunt for whales to conduct experiments with his diving machine and record observations in a journal.

On May 21, 1816—a calm but foggy day with no "fish" in sight—Scoresby tied together all the lead lines he could find onboard his

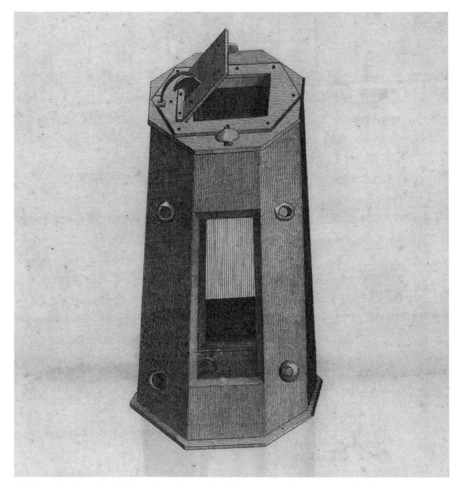

FIGURE 6.3. Scoresby's "Marine Diver." (William Scoresby, *Account of the Arctic Regions* [Edinburgh, 1820]; Courtesy of the Rare Book & Manuscript Library, University of Illinois at Urbana-Champaign.)

whaler, amounting to "somewhat more than 8/10 of an English mile" in length. He then attached wood blocks of different varieties to the end to test water pressure, and placed a thermometer inside the marine diver for temperature readings. Attaching the lead lines to his machine enabled Scoresby to lower the device to a depth of 600 fathoms. After some hours' experimentation, Scoresby observed that the maximum temperature of the water through which the marine diver passed lay

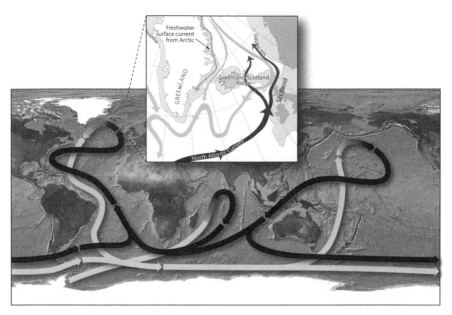

FIGURE 6.4. This schematized diagram of the global thermohaline circulation includes a dramatic U-turn in the North Atlantic in the vicinity of the undersea Greenland-Scotland Ridge. In the Greenland Sea west of Spitsbergen, where William Scoresby sailed in 1816, millions of gallons of warm, salty water flow northward, while the overturning southward flow is less salty and more than 10°F cooler. (Jack Cook; ©Woods Hole Oceanographic Institution.)

well beneath the surface. Also remarkable was the fact that this maximum temperature, 37°F, was the "greatest heat of the water, which [I] have observed in these regions." The experiment, concluded Scoresby, "proves the existence of a current from the southward running beneath, at the same time the current from the NE to the SW runs upon and near the surface, whereby the whole body of the polar ice is carried."[22] What Scoresby had unwittingly identified was, in fact, the main engine of northern hemispheric climate.

The so-called Atlantic Meridional Overturning Circulation (AMOC) is a submarine current system that transports tropical warmth to the North Pole via the gulf stream and, in the course of its many thousand miles' journey, moderates extremes of air temperature at all latitudes. The AMOC, in turn, belongs to the conveyor belt of thermal deep sea currents that girdle the globe from pole to pole. As warm waters flow into the North Atlantic, they grow colder and saltier, and hence more dense. The heavy water sinks, drawing lighter, deep water to the surface. With his diving machine, Scoresby was able to capture a snapshot image of this dynamic process of liquid thermal exchange, and a vivid impression of opposing southward and northward currents. Becalmed off the Greenland coast west of Spitsbergen (Svalbard), he was well situated to observe the AMOC in the deep polar basins of the Norwegian and Greenland seas where its overturning motion is in fact triggered. While the sinking northward current brought heat to the Arctic seas, melting the ice pack, the cold surface drift escorted the broken ice south into the Atlantic Ocean, wreaking havoc on transatlantic shipping lanes in 1816 and 1817. Because of his vast experience in northern waters, Scoresby was thus able not only to observe the dynamics of an overturning oceanic current funneling warm water north into the polar sea but to make note of its extreme behavior in that season.

With the redoubtable Scoresby's marine diver experiment, we inch closer to the solution to why, in the summers of 1816 and 1817—while the inhabitants of temperate zones from China to New England shivered and starved—the Arctic Circle basked in relative warmth and shed its ice at amazing, unprecedented rates. Crucial to understanding the relation between Tambora's eruption in 1815 and the reports of massive polar ice loss in the summers following is Scoresby's observation that, in 1817, the Arctic water was at its "greatest heat" in his experience. A survey of ships' logs in the period suggests environmental strains on the Arctic even beyond Scoresby's reckoning. Godthaab in southern Greenland experienced temperatures 5.5°C above average during the entire volcanic decade of 1810–19.[23] The record air temperatures in Greenland, as well as Scoresby's observation of unusually heated water, suggest that in the aftermath of Tambora, the AMOC was operating with increased intensity.

Answers to why this should be so may be found in studies of the 1991 eruption of Mount Pinatubo in the Philippines. The Pinatubo eruption has served scientists well as a model from which environmental impacts of the nearby Tambora event, unobserved by modern scientific

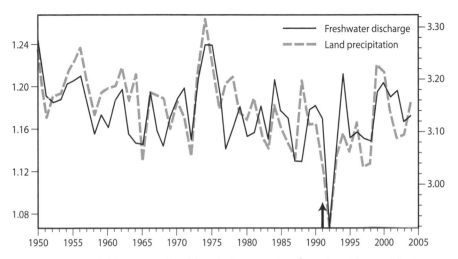

FIGURE 6.5. A model incorporating historical streamflow records of the world's largest 925 rivers shows the dramatic decrease in freshwater runoff to the oceans following Pinatubo's eruption in June 1991. (Kevin Trenberth and Aiguo Dai, "Effects of the Mount Pinatubo Volcanic Eruption on the Hydrological Cycle as an Analog of Geoengineering," *Geophysical Research Letters* 34 [2007]: L15702; © American Geophysical Union.)

instruments, might be extrapolated. A notable consequence of Pinatubo's eruption, and the global cooling it produced, was the "substantial decrease" in rainfall overland for a year following the eruption and a subsequent "record decrease" in runoff to the oceans. The cause was the chilled, volcanic atmosphere, which repressed evaporation and reduced the amount of water vapor in the air. Put in its broadest terms, reduced solar radiation in Pinatubo's aftermath altered the flow of energy through the coupled ocean-atmosphere system, with significant implications for the global hydrological cycle. Accordingly, the first post-Pinatubo year, 1992, witnessed the largest recorded percentage of the global landmass suffering drought conditions. A recent computer simulation of the influence of volcanic activity on global climate since 1600 produced the same "general precipitation decrease" in the high latitudes of the northern hemisphere, especially pronounced over land.[24]

In the case of Tambora, a volcanic event six times the magnitude of Pinatubo, hydrological disruption at the hemispheric scale must have been nothing short of catastrophic. In 1816 and 1817, with extreme drought conditions prevailing across the high North American land-mass, the Atlantic Ocean received only a fraction of its standard allot-ment of warm freshwater discharge from rivers and streams. As a result, surface waters in the North Atlantic became colder and saltier, sinking with greater force. The subsequent destabilization of the water column in turn enhanced the motive energy of the Atlantic thermohaline cir-culation. Convective currents released increased quantities of heat into the Arctic Circle, melting the ice cap, while a bulked-up southward cur-rent delivered great volumes of Greenland glacial ice into the Atlantic. The increased surface temperatures likewise inhibited the formation of new ice in the subpolar region, hence the magical-seeming open seas visible from the mastheads of British whalers off the coast of Greenland in 1816 and 1817.[25]

Volcanic enhancement of the AMOC is produced not only by re-duced air temperature, which suppresses evaporation. Wind also plays its part in draining water vapor and energizing ocean circulation. As we saw in chapter 3, gales swept across the North Atlantic with unusual force and frequency in the Tambora period. A major tropical eruption enhances the normal gradations in temperature between the equator and the poles. Differentiated temperatures in turn influence the den-sity and pressure gradients that power winds, strengthening wafts into breezes and stiff breezes into gales. For the North Pole region, this meant an amplified, positive phase of the Artic Oscillation, its major circulatory weather system. These stronger-than-usual winds further cooled the sea-surface temperatures of the North Atlantic, adding to the positive inputs at work on the AMOC. In short, environmental change in the Arctic in the Tambora period was driven mostly by changes in oceanic currents and winds, which overrode the general atmospheric cooling of the planet. And because wind and ocean currents behave nonlinearly in response to atmospheric change, the Arctic ice pack was vulnerable in turn to extreme transformations.

It is important to note how regionally specific the strange phenom-enon of volcanic heating of the Arctic was (and is). Anywhere outside the warming embrace of a steroidal AMOC, it was cold, cold, cold. The Hudson Bay in 1816 and 1817, for example, in the Canadian subpolar region, witnessed its coldest temperatures and greatest ice extent in 120 years of record keeping.[26] The trading ships of the Hudson Bay Company, on their annual voyage from England for precious Canadian furs, met with an impenetrable peninsula of ice past the Hudson Strait, drifting helplessly for months at the mercy of the pack. To the north, however, the weather, as Kotzebue found, was eerily "delightful." Thanks to Tambora—grand saboteur of the global climate system—the seaward door across the top of the world stood tantalizingly open.

THE MAN WHO ATE HIS BOOTS

Because the atmospheric residence of volcanic sulfate aerosols—even those of a major tropical eruption—is no more than three years, John Barrow's dreams of an easy passage across an ice-free Arctic, not to mention balmy English winters, never materialized. The polar north warmed dramatically in the years 1816–18 before just as suddenly entering a renewed cold phase. A baffling and tragic irony, then, that these same cold decades witnessed the most concerted British assault on the Arctic.

So confident was Barrow in the first polar missions of 1818 that he directed Captains Ross and Buchan to rendezvous in the Pacific after completing their pleasure cruise across the Arctic Circle. But Buchan's mission to the east of Greenland quickly came to grief on the massive ice pack north of Spitsbergen. Not equipped with sleds for an overland bid at the pole (as William Scoresby had advised), Buchan had no alternative but to turn back with nothing to show for his efforts. In this case, the arch spinmeister Barrow chose silence as the best public relations policy. A written account of the failed Buchan expedition didn't appear until the 1840s.

Barrow reserved a more aggressive reception for John Ross, who returned early from his expedition to Baffin Bay to declare that he had encountered a range of mountains to the west of Lancaster Sound, terminating any northwest route across the Arctic. Barrow was crushed. But when rumors reached him that other officers in the expedition—

notably Edward Parry, who commanded Ross's companion ship—by no means concurred with their leader's view, and that the "mountains" were likely the figment of a frostbitten captain's imagination, he launched into print with a withering assault on Ross's competence and courage.

The clouds of 1818 contained a silver lining for Secretary Barrow. The embarrassing Ross expedition had turned up a likely hero in the form of Lieutenant Parry, whom Barrow immediately entrusted with a follow-up expedition for the summer of 1819. Sir Joseph Banks himself entertained the dashing young officer in his study, pulling out a newly drawn map that showed the vast Greenland seas empty of ice (Scoresby's reports had quickly become official geography). Positive results were immediate. Parry's ships sailed blithely *through* Ross's mountain range named, ever so briefly, "Croker's Mountains," after Barrow's immediate superior at the Admiralty (the same baleful Croker who published a scathing review of *Frankenstein* in the *Quarterly Review* later that year). Passing through Lancaster Sound, Parry encountered encouragingly open waters, and his two ships reached as far as Melville Island at 110°W by summer's end, the farthest point west ever reached.

With a generous Admiralty reward already earned for their extravagant longitude and dreams of completing the northwest passage the following year, Parry made the critical decision to spend the sunless winter months on the Arctic ice. A superb officer and "man-manager," Parry kept spirits high through the long months of darkness with a lively schedule of games and theatricals. He even founded a polar newspaper, with dubiously witty contributions from the officers copied out in journal format, one for each ship. Parry never made it any farther toward his Pacific destination, but the homosocial romance of his winter quarters—his happy, orderly Little-Britain-on-Ice—delighted the public and made him an instant celebrity. Parry of the Arctic embodied a new leadership ideal for postwar Britain: virile, liberal, and humane—just the sort of "new man" Jane Austen's heroines fall for in her novels published that same decade.

Most important, Parry's success had secured a generous stock of credit for Barrow's long-term polar enterprise—a deep public goodwill

FIGURE 6.6. A dashing portrait of Captain William Edward Parry on his triumphant return to England, having reached Melville Island and over-wintered in the Arctic in 1819–20, the first expedition ever to achieve these feats. Parry's first voyage stands as the greatest unqualified success in the history of nineteenth-century British polar exploration. When, in 1826, the Bronte children received a box of toy soldiers around which they were to build their elaborate fictional kingdom, eight-year-old Emily, future author of *Wuthering Heights*, named her favorite "Edward Parry." (© National Portrait Gallery, London.)

that, despite repeated disappointments, would not be fully exhausted until after Barrow himself had died of old age. Long before then, however, both Barrow's influence on the narrative of Arctic exploration—and that of his hero-designate Parry—had been eclipsed by the iconic figure of John Franklin, "The Man Who Ate His Boots."

Barrow arranged Franklin's first expedition on the cheap. Critically under-resourced, and with no experience in Arctic conditions, Franklin's overland trek across the Canadian north in 1819–22 quickly descended into a mobile purgatory of starvation, insanity, murder, and cannibalism. Franklin and his ever-dwindling band of men wandered the snowy tundra for months on end like living skeletons, subsisting on mossy lichen and burnt leather. Whatever distress the nation felt at the news of Franklin's Arctic disaster, however, quickly gave way to pride. At last, in Franklin, the nation might welcome a worthy naval successor to Lord Nelson. After all, the man had eaten his own boots for king and country. Franklin's wife, the poet Eleanor Porden, had published some jingoistic verses on the eve of his first expedition with Buchan. As she eerily foresaw in her 1818 poem "The Arctic Expeditions," her beloved Franklin would "furnish tales for many a winter night" for the British public to feed upon "with strange delight."[27] When presented with the choice between Parry's wholesome sociability on the ice and the horrors of Franklin, the people chose horror.

When Franklin returned to the Arctic a final time in 1845, he commanded two vessels laden with tinned food, fine china, chandeliers, and a complete gentleman's library. The Admiralty wished, it seemed, to make amends for the scant provisioning of his earlier tour. This time, he would preside over a miniature maritime empire while floating serenely through the northwest passage. But in Lancaster Sound—where, a quarter-century earlier, Parry had sailed clear through Croker's Mountains—Franklin's expedition ran into a wall of ice. Hopelessly trapped and disoriented, they eventually abandoned their ships. Of the 128 crew on Franklin's mission, not a single man returned. The indigenous people of the Arctic looked on in wonderment and sorrow as bands of ghostlike men wandered over the ice, hauling their china and books with them. Some succumbed to lead poisoning from

Auffindung des Franklin'schen Bootes auf König Williams Land.

Zu Franklin's Expedition.   Leipzig: Verlag von Otto Spamer.

FIGURE 6.7. The grisly discovery of skeletal remains of Franklin's crew on King William Island, by the McClintock search expedition in 1859, is illustrated here in a German volume from 1861. Accounts of the British polar voyages were translated and published all across Europe, including Norway, where they inspired the young Roald Amundsen. (Hermann Wagner, *Die Franklin-Expedition und ihr Ausgang* [Leipzig, 1861], 218; Courtesy of the Rare Book & Manuscript Library, University of Illinois at Urbana-Champaign.)

their tinned provisions, the rest to hypothermia and starvation. In their extremity, they resorted to feeding on their dead messmates. Franklin himself was spared the worst, being one of the first to die.[28]

With Franklin's death, the objectives of polar exploration underwent a sea change, from a scientific inquiry into the northwest passage and its commercial prospects for the British empire to a morbid quest for Franklin's salvation, or his remains. On one of Parry's unsuccessful return trips to the Arctic in the 1820s, he had difficulty explaining to the Inuit inhabitants he met of the purpose of his voyage. The tribespeople decided among themselves that he could only be in search of his ancestors' bones. Nothing else made sense.[29] Franklin wasn't yet dead, but their judgment proved unerringly prophetic. When Robert McClure, during one mission for Franklin's recovery, actually mapped the course of the northwest passage, putting an end to the centuries of speculation and desire, his achievement received muted treatment in the press. It is a statue of Franklin, not McClure, that stands in central London, with its famous inscription by Tennyson:

> Not here! The white north hath thy bones, and thou,
>   Heroic sailor soul,
> Art passing on thy happier voyage now
>   Towards no earthly pole.

Only according to the gothic metrics of Victorian Arctic romance could Franklin, who only ever failed, outdo the brilliant and resourceful Parry by virtue solely of his body count, a victory certified by his own martyrdom. Barrow's optimistic 1818 proclamation of a new golden age of global warming, and an open polar sea, had long been forgotten. The polar seas had opened briefly in the Tambora period—just enough for Parry's 1819 expedition to raise the nation's hopes for completion of the fabled northwest passage to the Pacific. But post-1819, with North Atlantic ocean circulation returned to normal, the polar ice abruptly closed over once more, like a grave.

Mary Shelley's *Frankenstein* stands for so much in Western culture—hubris, horror, and schlock—that it's easy to forget the novel also contains the first significant statement of polar skepticism in the nineteenth century. Like most educated Europeans of her day, Shelley was addicted to travel literature. She had been reading anthologies of old voyages for pleasure and was preparing her own Alpine travel journal for publication when she came across Barrow's *Quarterly Review* article calling for a new generation of polar explorers to complete the quest for a northwest passage.[30]

After two and a half years living with Percy Shelley, Mary had a gimlet eye for dangerous romantic excess in men. She saw it plainly in the "boy's own" Arctic propaganda of John Barrow. Barrow's roll call of patriotic polar adventurers, for "men zealous for their country's weal, and the honour of science," inspired in Mary the idea of an Arctic frame narrative for her novel.[31] The opening and closing chapters of *Frankenstein* feature an idealistic but inept polar explorer named Walton who rescues Frankenstein, his doomed alter ego, on the frozen northern wastes. His role in the novel is to bear sympathetic witness to Frankenstein's death and to record his tragic history. If Barrow intended his polar journalism to rouse the spirit of masculine adventure and national ambition in his readers, he thus failed utterly with Mary Shelley. Instead, through the figure of Walton, she turned a laser-like skepticism on Barrow's windy mythology of the Arctic, in which she detected the same hubris and reckless disregard for human costs that characterize her protagonist Frankenstein.

Evidence from the text of *Frankenstein* suggests that Mary found material for the Walton character in one passage from the *Quarterly Review* article in particular, in which Barrow mockingly recounts the misadventures of a low-ranked naval officer named Duncan who, in 1790, set out to explore the northern reaches of Hudson Bay in Canada. "Never," wrote Barrow, "was man more sanguine of success in any undertaking than Mr. Duncan." *Frankenstein*, likewise set in the 1790s, correspondingly opens with Walton's airy optimism: he dreams he will discover an open Arctic sea "where frost and snow are banished . . . a land surpassing in wonders and in beauty every region hitherto discovered on the habitable globe . . . a perpetual splendour." But Shelley's Walton, like his historical counterpart Duncan, soon finds himself beset by ice, misfortune, and a mutinous crew. Barrow's contemptuous description of Duncan's failed efforts—"grief and vexation so preyed on his mind as to render a voyage which promised every thing, completely abortive"—in turn offered Mary Shelley the perfect model for Walton's shame at the conclusion of *Frankenstein*, when he abandons his quest for the Edenic North Pole of his imagination, acutely "disappointed" and quasi-suicidal.[32]

In the concluding drama surrounding the abandonment of Walton's polar quest, Shelley's Dr. Frankenstein is made to sound like John Barrow himself. In an impassioned speech he delivers to Walton's crew,

Frankenstein implores them not to abandon their "glorious expedition" but to "return as heroes who have fought and conquered" the terrors of the Arctic. His patriotic rhetoric, like Barrow's on the British public, has a spellbinding effect: "when he speaks, they no longer despair; he rouses their energies, and, while they hear his voice, they believe these vast mountains of ice are mole-hills, which will vanish before the resolutions of man."[33] For the reader, however, Frankenstein's speech, like his scientific idealisms, rings hollow. We take leave of Shelley's novel persuaded that the romantic quest for a northwest passage stands second only to the reawakening of corpses with electricity as an example of extreme human folly. Her friend and fellow polar skeptic, Lord Byron, put it with customary pith:

> . . . voyages to the Poles,
> Are ways to benefit mankind, as true,
> Perhaps, as shooting them at Waterloo.[34]

It was to be many decades before Mary Shelley's Arctic warnings were heeded—decades of frostbite, starvation, lost ships, and lost men. At an early point, public fascination with the polar quest became untethered from any worldly measure of success. Instead, the enterprise took on the characteristics of a neo-Arthurian cult, to which Britain's finest knights would naturally be sacrificed in search of the elusive grail. Barrow's mythology of the northwest passage took hold of the public imagination in ways beyond his wildest dreams. Thousands of newspaper column inches chronicled the British polar expeditions between 1818 and 1860, not to mention a veritable blizzard of stories, plays, panoramas, songs, political speeches, paintings, prints, and photographs. The Arctic explorers themselves—Parry, Ross, Franklin, and their suffering bands of brothers—enjoyed massive celebrity: the narratives of their bleak, often nightmarish journeys were devoured by millions worldwide. The bitter, gothic romance of polar exploration, first to the Arctic then to Antarctica, evolved into a defining cultural symbol of the Victorian period in Britain.

Given the dozens of history books devoted to the significance of British polar exploration, how poignant then to learn that none of it might have occurred at all—or, at least, that the polar history of the nineteenth century would have taken a vastly different course—were it not for the eruption of Mount Tambora in April 1815, which wrought temporary but radical environmental changes on the Arctic Circle. Tambora's volcanic dust might have vanished from the atmosphere a few years after 1815, but the Edenic prospect it opened—of an unfrozen north of "perpetual splendour"—persisted a full century in restless British fancies. Only the definitive failure of Robert Scott in the Antarctic in 1912—coupled with the triumphant expeditions of Roald Amundsen who claimed the Northwest Passage and both poles for Norway—at last drove a stake through the dead heart of John Barrow and his fever dreams of British polar dominion. The subsequent trauma of World War I—which put an end to Victorian fantasies of many kinds—ensured that this time there would be no full-fledged revival, no new Franklin (or Frankenstein) to sacrifice himself gloriously on the ice.

CHAPTER SEVEN

ICE TSUNAMI IN THE ALPS

While John Barrow peddled his theories of global warming to a gullible Admiralty in late 1817, others put a more commonsensical gloss on the cold, violent weather of the Tambora period. The world was getting permanently colder across the entire hemisphere: a frightening age of glaciation was underway. In a long essay baldly titled "Climate," a writer for the *Morning Chronicle* in London reflected gloomily on the deteriorating atmosphere:

> In America, as well as in Europe, the climate and temperature of the air seem to have undergone an equal vicissitude within the last few years. The changes are more frequent, and the heat of the sun is not so early or so strongly experienced as formerly . . . verify[ing] the theory of those observers of nature, who have said that the extreme cold of the north is gradually making encroachments upon the extreme heat of the south.

Contra Barrow, this writer interprets the increased presence of ice floes in southerly latitudes as evidence that the polar ice cap is rapidly *expanding*, not breaking up. In addition to the authority of reliable "observers of nature," the author draws his readers' attention to "authentic reports of the best informed travellers" to the Alps, where the glaciers "continue perpetually to increase in bulk." An obvious connection must exist, he argues, between the increase in polar ice and the menacing advance of Alpine glaciers.

The *Chronicle*'s climate prognosticator fears the worst and concludes his article with a general appeal to the authorities to "make every effort, to which human ingenuity and strength are competent . . . for the purpose of counteracting the growing evil." Because he views the deteriorating climate in hemispheric terms, the writer advocates international cooperation. As a starting point, he suggests, the navies of the world might usefully combine to "navigat[e] these immense masses of ice into the more southern oceans."[1] While John Barrow dreamed of sending his heroic officers on a northward cruise into balmy polar seas in 1817, our author charts a diametrically opposite course for the British navy. Their post-Napoleonic mission? To chaperone titanical icebergs in the direction of the tropics.

"THE RACE OF MAN FLIES FAR IN DREAD"

"The best informed travellers" the London journalist quotes regarding Swiss glaciers in 1817 could not have included the obscure continental tourists Mary and Percy Shelley. But the Shelleys, on a tour of the Alps in July 1816, were keen eyewitnesses to the alarming glaciation of the Tambora period, and their imaginative reflections on the subject have long outlasted those of the *Morning Chronicle*'s sources. In the summer of 1816, the Shelleys entered an Alpine landscape in the throes of atmospheric cooling wrought by Tambora's eruption half a world away. As we have seen, the average Swiss temperatures for the summer months that year reached historic lows, around 14°C. The summer's maximum "warmth" in July and August was barely that of a very cold June day in today's terms. Moreover, these months were extremely wet, making life miserable for both villagers and tourists, and smothering the mountaintops in record snows. Accustomed to the summer pasturing of their cattle on the Alpine meadows, the highland farmers could only look on anxiously as the low winter snow line persisted into the spring.

Huge avalanches, normally a feature of the springtime melt, continued through August.[2]

The Shelleys observed at close quarters the ecological impacts of the volcanic summer of 1816 across the Alpine landscape. Riding mules for the ascent to the tourist hub of Chamonix, they barely negotiated a raging torrent that, three days before, had "descended from the snow & torn the road away." The ever-present clouds of that summer obscured the airy summit of Mont Blanc as they passed into the valley of Chamonix, but this minor disappointment was soon assuaged by another highly gratifying expression of the mountains' elemental power:

> Suddenly we heard a sound as of a burst of smothered thunder rolling above.... Our guide hastily pointed out to us a part of the mountain opposite from whence the sound came.—It was an avalanche. We saw the smoke of its path among the rocks & continued to hear at intervals the bursting of its fall—It fell on the bed of a torrent which it displaced & presently we saw the torrent also spread itself over the ravine.[3]

The thunderous avalanche signaled a glacier on the march. The smoky deluge of snow, soon to harden into ice, marked a glacier's first claim upon its extended frontier. Glaciers are "sensitive barometers of climate change," capable of rapid and dramatic response to fluctuations in atmospheric temperatures and precipitation. Summer temperatures are especially decisive, accounting for 90% of interannual growth in glacier mass.[4] As the Shelley party would soon learn, the Alpine ice had been particularly aggressive that year, locking up valuable highland pastures and advancing menacingly upon the populated valleys. The empire-hungry Napoleon, who had in 1800 marched his armies across these mountains to conquer Italy, now languished in exile, but the cold, snowy years of 1815–18 had stirred the territorial ambitions of an even greater force of nature.

The literary members of the Shelley Circle, for all their differences in style, shared the talent of every serious writer: they transformed their ordinary lived experiences into art. Descriptions of events in their personal letters very often show up, in altered form, in their poems or fiction. So it was with the Shelleys' tour of the Mont Blanc glaciers in the cold summer of 1816. The night of their arrival in Chamonix, Shelley described in a letter to his friend Peacock in London his exhilarated wonder on first coming into view of Europe's highest mountain range. "I never knew I never imagined what mountains were before," he wrote, groping for words. "The immensity of these aerial summits excited, when suddenly they burst upon the sight, a sentiment of extactic [*sic*] wonder, not unallied to madness."[5] The next day, housebound once again by rain, Percy revisited the experience in verse form as the poem "Mont Blanc," with its famous concluding challenge to the Alpine summit:

> And what were thou, and earth, and stars, and sea,
> If to the human mind's imaginings
> Silence and solitude were vacancy?

In other words, the highest peak in Europe would be nothing without Shelley there to exalt it.

Percy's day trip to the glacier of Bossons, which skirts the southern slopes of Mont Blanc, had produced conflicting emotions. He was struck, as all Alpine tourists were, by the immense rivers of ice descending from the mountains to the valleys and by the contrast between "the green meadows & the dark woods with the dazzling whiteness of its precipices and pinnacles." But he seems to have been genuinely appalled by his guide's information that the glacier was *advancing* at the rate of a foot a day over the valley, swallowing up the pretty Swiss landscape like a giant white python: "These glaciers flow perpetually into the valley," Shelley wrote that evening, "ravaging in their slow but irresistible progress the pastures & the forests which surround them, & performing a work of desolation in ages which a river of lava might accomplish in an hour, but far more irretrievably." Poet-scientist that he was, Shelley could look over the white stillness of the Bossons glacier and intuit within it the rage of a volcano, like the slopes of Vesuvius he would climb two summers later. Shelley had read educated accounts of the glaciers that theorized their perpetual advance and decay, but his guide assured him that the local people held a darker view—namely, that the glaciers would eventually smother the entire valley, as they had done in the ancient times.

The French-American travel writer Louis Simond, traveling through the same valley the following summer, observed that the winter of 1816–17 had been "remarkably mild" around Chamonix but that snowfall levels on the mountains were tremendous. As a result, the valley's twin glaciers—the Bossons and the Mer de Glace—had advanced more than a hundred feet beyond their usual range. He gives a vivid account of the glacier's slow-motion destruction in terms reminiscent of Shelley's "Mont Blanc":

> With slow, but irresistible power, the ice pushes forward vast heaps of stones, bends down large trees to the earth, and gradually passes over them. . . . Streams of water of a milky appearance, continually issuing from under the glacier, formed new channels through the adjacent meadows.

These meadows represented cultivated land, dotted with valuable farmhouses, barns, and mills, whose icy submersion the local farmers were powerless to prevent. Simond witnessed the quiet unfolding of this human tragedy in the Chamonix valley in 1817: "The miserable inhabitants, collected into melancholy groups, looked on dejectedly. . . . Several dwellings are actually under the glaciers, and others await the same destruction."[6]

The dejection of the montagnards of Chamonix is understandable. The climate historian Christian Pfister has established a triangular correlation between cold summers, glacial advance, and historical peaks in grain prices in Switzerland.[7] For the Alpine peasantry, energized glaciers signified the risk of starvation, not merely an interesting geological phenomenon. By the end of the Tamboran cooling regime in 1818, the Bossons glacier had submerged some five hectares of farmland in the valley and threatened the village of Monquart. The hapless residents, taking a providential view of the glacier's expansion, performed a ceremony of appeasement in which they planted a large cross at its voracious rim.[8]

The prospect of a greater glaciated Alps upset Shelley, whose feelings of excited wonder on first entering the valley now gave way to creeping dread. Observing the lines of forest trees flattened by the encroaching tongue of the Bossons glacier, Shelley found "something inexpressibly dreadful in the aspect of the few branchless trunks which nearest to the ice rifts still stand in uprooted soil. The meadows perish overwhelmed with sand & stones. Within this last year these glaciers have advanced three hundred feet into the valley."[9] The sight of the expanding glacier brought still another reference to Shelley's encyclopedic mind. The esteemed French naturalist the Comte de Buffon had proposed that "this earth which we inhabit will at some future period be changed into a mass of frost"—in other words, an Ice Age, though it would be another two decades before that term—foundational to modern climate science—was invented. To trace the origins of Ice Age theory to Tambora-era glaciation is the purpose of this chapter.

Shelley's intuition of Alpine danger in 1816 should be tied to the specific climatic conditions of that cold Tamboran summer, in which the glaciers neared their modern historical maximum and appeared grimly unstoppable: "The race/Of man, flies far in dread," he wrote. But what Shelley could not have known was that his fears, in the geological short term, were needless. The icy augmentation Shelley was witness to, wrought by Tambora's chilling hand, represented the southernmost glacial extent of the Little Ice Age. The glaciers of the Mont Blanc massif would never again extend so far as they did for the Shelleys in the Tamboran period.[10] Two centuries later, our perception of an Alpine crisis is, of course, very different. With the accelerated global warming of the past half century, Alpine glaciers are now in unprecedentedly rapid retreat, raising the very real prospect of an ice-free Europe by century's end, with disastrous implications for continent-wide water systems and agriculture.

Meanwhile, back at the hotel in Chamonix, Mary Shelley's imagination was far from idle. She considered the Alps "the most desolate place

FIGURE 7.1. J.M.W. Turner's etching from 1812 shows Mary Shelley's view from the valley of Chamonix toward the Mer de Glace. The longest glacier in France, it courses down the northern slopes of the Mont Blanc range. Turner's image highlights magnificently the aquatic instability—and grim threatfulness—of the "Sea of Ice." (© Fine Arts Museums of San Francisco/Achenbach Foundation.)

in the world," which, coming from a true-born Romantic, was high praise indeed.[11] Their walking tour of Mont Blanc's northside glacier, the Mer de Glace, had put her in mind of the ghost story competition with Byron and Shelley. She was inspired to "write my story," as she confided to her journal, and now had a setting for the opening of book 2 where she would reunite the unfortunate monster, lately a murderer on the run, with his unhappy creator. In one of the most gripping scenes of the novel, Frankenstein embarks on his Alpine tour with the hope of discovering in the majestic stillness of the mountains some escape from his grief and regret: "My heart, which was before sorrowful, now swelled with something like joy." Instead, he finds that the glaciers have become the refuge of his hideous Creature, who waylays him on the vast tundra of the Mer de Glace: "I suddenly beheld the figure of a man, advancing toward me with superhuman speed. He bounded over the crevices in the ice."[12] Mary Shelley's staged gothic encounter on the Alps in *Frankenstein* echoes Percy's ambivalence in his poem "Mont Blanc." Both detect in the sublime beauty of the glaciated peaks an undercurrent of horror.

The Shelleys were not alone in their dread of advancing Alpine glaciers in the waning years of the 1810s. The members of the new Swiss Society for Natural Science, alarmed that the widespread glacial growth of that decade had accelerated markedly on account of the recent cold summers, turned their attention to glaciology for the first time in their annual meeting in Bern in 1816. The following year they announced a new competition. A prize was to be awarded to the scientist who conclusively answered the question of whether the climate of the Swiss Alps had grown colder. In addition, applicants were required to provide their observations on the growth and decay of glaciers. Three years later, however, the society had received a total of one mediocre submission, a testament not to lack of interest in the question but to its difficulty.

Enter Ignace Venetz, a gifted young engineer and montagnard from the Swiss canton of Valais, who followed the Shelleys' footsteps along the tourist trails of Mont Blanc in the summer of 1820.[13] His impression of a glaciated Alps was as powerful as Shelley's, but his educated eye was better able to decode their ice-bound history through the specific evidence of moraines. Buried beneath swaths of forest, or winding sinuously miles from the present glacial rim and snow line, these riverine piles of rock and earthy debris marked the ghostly outline of the massive glaciers of the past. In his seminal paper of the following year, "Mémoire sur les variations de la température dans les Alpes de la Suisse," delivered to the Swiss Society, Venetz laid down the founding principles of Ice Age theory. Only historical variations in temperature could explain the changes in glacial extent, he argued. Furthermore, these changes in climate were extreme enough to have once submerged this region of the Swiss Alps beneath a vast sheet of ice, whose subsequent retreat under warmer temperatures had left behind the moraines and striated rocks as a kind of glaciological signature.

The theoretical possibility of large-scale glaciation had been suggested intermittently over the previous decades by scientists in various

FIGURE 7.2. Portrait of Ignace Venetz (1826). The artist has ingeniously incorporated Venetz's most famous engineering achievement—the tunnel at the Giétro glacial dam—as the background Alpine vista. (© Musées cantonaux, Sion; Photo: François Lambiel.)

countries, only to be ignored. Venetz was the first to make a serious case based on geological observation, and the first to make the necessary link between glaciation and climate change. Venetz's career as a pioneering glaciologist was marked by a lifelong devotion to on-site study in the Alps, but also frustratingly few publications. It is possible, nevertheless, to show that the evolution of his theory of the Ice Age, and its breakthrough revelation, occurred in the course of a specific Swiss crisis during the Tamboran summers of 1816–18, when the multiplying spin-offs of sustained climate deterioration threatened lives and livelihoods across the Alps.

DOOMSDAY IN THE VAL DE BAGNES

Two years after Shelleys' tour of Mont Blanc, in the spring of 1818, the residents of the Val de Bagnes, forty kilometers to the east along the same mountain range, confronted a chilling sight. The River Dranse, their major water source, which ought to have been swollen with spring meltwater from the mountains, was reduced to a trickle.[14] Villagers sent to investigate along the slopes of Mount Le Pleureur, where the Dranse flowed through a narrow gorge at the head of the valley, returned with grim news. The decade's succession of cold summers, more intense since 1815, had left a potentially disastrous legacy in the form of an ice dam created by the Giétro glacier. Its lip advanced to the very brink of the narrow gorge, the Giétro had begun to deposit huge blocks of ice into the river, forming a cone-shaped dam thirty meters high. A huge lake had now formed behind the icy wall: three and a half kilometers long, two hundred meters wide, and up to sixty meters deep.

Periodic buildup and release of large amounts of water are a natural characteristic of glacial systems, and typically occur in the spring after a cold summer when meltwater coursing from the peaks and slopes, then draining into a river system, meets with an unusually resilient barricade of ice.[15] As the weather warms and the volume of water builds, the pressure on the dam increases to a breaking point. In Iceland, the ensuing catastrophic floods are called *jökulhlaups*, meaning "glacier-floods." The French word *débâcles* is a more emotive term (at least for English speakers) for the same phenomenon, conveying a sense of their devastating impact on vulnerable Alpine communities.

Extreme glacier-flood conditions prevailed across the Alps in the spring of 1818. In the rising midyear temperatures, and with its volume

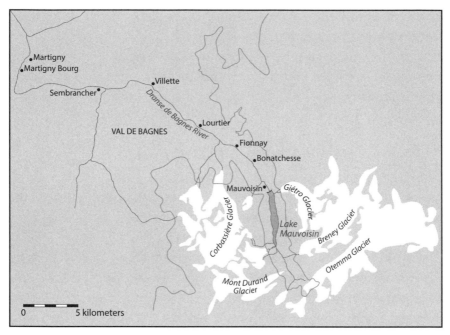

FIGURE 7.3. Map showing the concentration of glaciers around Mauvoisin. The mountainous borders of the Val de Bagnes created a funnel for the pent-up waters of the Dranse in 1818, with the market town of Martigny in its direct path. (Jean Grove, *The Little Ice Age* [London: Methuen, 1988], 174.)

ever-increasing from seasonal runoff, the dammed-up Dranse was poised to burst at any moment. The threatened flood would inundate the pristine valley with over twenty million cubic meters of water, enough to submerge the town of Martigny twenty kilometers away, and destroy all the farmland and villages in its path. The worried peasants of the Val de Bagnes were fortunate in one regard, however. The provincial government had sent Venetz—a true montagnard—to deal with the crisis. In return, the valley residents offered Venetz their most knowledgeable guide, a chamois hunter from the town of Lourtier named Jean-Pierre Perraudin, to accompany him on his tour of the site of potential cataclysm.

Perraudin is a remarkable figure in the history of nineteenth-century science.[16] With little education and no academic credentials whatever, he nevertheless took it upon himself to convert any science-minded individual who crossed his path to the long-held local belief that the

Val de Bagnes had once been covered in a vast sea of ice. For evidence of this, he pointed to the existence of striated marks on rocks high above water level; to the presence of moraines that seemed to mark the outlines of an enormous ancient snake of ice now vanished; and to the anomalous, high-up location of giant boulders, called "erratics," whose mineral constitution did not match that of the rock formations around them.

Joining Venetz and Perraudin on their urgent trek to the Giétro glacier was the director of the nearby salt mines at Bex, Jean de Charpentier, a respected naturalist whom Perraudin had already attempted to convert to his glacial theories, without success.[17] Charpentier later recalled their geological conversations, in which he had found himself persuaded that the prevailing view of the transport of enormous boulders by water, especially uphill, was indeed an impossibility. Nevertheless, Perraudin's larger idea that the entire Rhône Valley had once been submerged beneath a sheet of ice hundreds of feet thick struck him as pure hogwash, a fantasy "so extravagant that I considered it not worth examining or even considering."[18]

Charpentier's reaction was typical of nineteenth-century geologists when first introduced to glacial theory: amused disbelief. The emerging scientific communities of Europe prided themselves on their intellectual sobriety—their quasi-sacred commitment to observed phenomena and empirically testable facts. To succumb to fantasies of an ice-bound planet—a world so utterly different from the visible one—spoke of medieval-style madness, a nonsense superstition. In key individual cases, however—including Venetz, Charpentier, and later Charpentier's student Louis Agassiz—this initial outright rejection of glacial theory was followed by a period of increasingly serious reflection on the evidence in its favor, culminating in a conversion to the glacialist cause almost religious in its intensity.

One imagines Perraudin teasing Venetz's curiosity with the heretical image of vast Alpine ice sheets as they clambered up Mount Le Pleureur toward the Giétro glacier in the spring of 1818. Perhaps Perraudin knew that Venetz, in addition to his day job as engineer of Valais, was an amateur naturalist with a specific interest in glaciers and had given a paper two years earlier before the Swiss Society of Natural Sciences on the subject of glacial advance and the transport of debris. This early paper had made no mention of large-scale glaciation, however, and did not challenge the established theory that the geological features of the Alps had been caused by aquatic submersion—by the very sort of catastrophic flood that now threatened the Val de Bagnes.

No doubt Jean de Charpentier would have listened in to the conversation and expressed his doubts. But Perraudin was persistent by nature and had reason on his side. How could these rocks be scarred, those moraines have formed, or these boulders have been transported miles from their place of origin except by the powers of glaciation? No flood could engineer earth removal at such a scale. One can imagine Charpentier and Venetz looking at each other and shaking their heads. Perhaps Perraudin's status as an uneducated peasant made it all the more difficult to accept what was, for the time, a truly outlandish notion.

But Perraudin had done his job. The seed of Ice Age theory was planted. It had crossed the threshold separating folk belief and educated, scientific opinion, and though it would take decades to emerge fully formed, the logic of glaciation was now as unstoppable as the mighty lake forming behind the ice dam on the River Dranse. These three Swiss men—Venetz, Perraudin, and Charpentier—represent a who's who of early glacial theory, and thus modern climate science. Their happenstance gathering in the mountains above the Val de Bagnes in spring 1818, however it precisely occurred, marks the first dynamic concatenation of glacialist ideas that would ultimately lead to the establishment of Ice Age theory. It was the explosion of a faraway mountain, Tambora, that first brought them together as a group to tackle one of that volcanic event's many spin-off threats, in this case an Alpine *jökulhlaup*.

What Venetz found on arrival at the gorge appalled him. For this vast dammed lake to empty itself into the Val de Bagnes was to imagine a debacle of truly biblical proportions. But with energy and ingenuity to burn—and deep pockets for payment of danger money to his workers—Venetz set about averting the crisis. First of all, he ordered the cutting of a tunnel through the giant cone-shaped ice wall in the hope of managing the gradual release of water from the dam. This task was carried out under the intermittent hail of enormous blocks of ice breaking from the glacier overhead, which sent large waves over the top of the wall of ice. Added to that was the ever-present, stomach-churning threat that the dam itself might suddenly give way, sweeping the entire party to their deaths in a calamitous torrent.

Making the situation more difficult still, the weather was terrible. Tambora was not yet done with the Alps. After two feet of snow fell in two days in mid-May, most of the workers quit in protest at the freezing conditions. They worked night and day, hacking at the ice with hatchets while soaked in icy water, wearing ordinary shoes instead of water-proof boots. Venetz sent the demoralized Italians home and lured his toughest Swiss montagnards back with the promise of bonuses. Work continued around the clock for a month as the water level of the dam continued to rise ominously toward the foot of the tunnel. Venetz himself camped out on the ice. Would the ice wall hold long enough for the tunnel to work its draining effect? Several false alarms led to premature evacuations of the valley. And when, on the evening of June 13, water began to issue through Venetz's tunnel, the state of high alert across the valley began to relax. Three days later, the water level of the dam had dropped by ten meters. By the time the dam ultimately burst, Venetz's tunnel had reduced its volume by a full third, sparing the valley a worse calamity.

Because Venetz could not quite save the day. The water coursing through the tunnel, combined with that continually cascading from the Giétro glacier above and through myriad other cracks in the ice, dramatically reduced the thickness of the wall until finally, at 4:30 in the afternoon of June 16, the frozen edifice suddenly collapsed with an enormous, deafening crash. Fifteen million cubic meters of heaving water, ice, and mud thundered down into the Val de Bagnes. Venetz and his workers scrambled to higher ground and watched in horror as the pent-up River Dranse, now a hundred-foot-high tsunami, rushed through the gorge, dragging boulders and great blocks of glacial ice with it before launching itself with a roar into the valley. The old stone bridge at Lake Mauvoisin was smashed in an instant. Dozens of hill-side chalets fell victim to the great onrushing tide, while in the valley farmhouses and barns bobbed on the torrent like giant toys. The flood ripped entire forests from their roots, as glossy orchards and fields of wheat and grain were submerged, including the family farmlands of Jean-Pierre Perraudin.

By the time it reached the plain, the flood had assumed a truly sinister character: an oozing lake of black mud filled with rocks and tree trunks churned toward the River Rhône at Martigny, accompanied by a thick, black fog. Vital infrastructure and industries of the valley—roads, bridges, sawmills, flour mills, and an ironworks—sank beneath the miasma. In the words of one eyewitness the entire valley, "but a moment before so beautiful and so populous, was converted in a moment into a dreary desert."[19] In Martigny, this wall of mud, ice, and dangerous debris flooded the streets and houses, reaching up to the second story.

The icy deluge took half an hour to pass. It spilled across the River Rhône, and only exhausted its destructive rage on reaching Lake Geneva around midnight, into whose vast depths it was finally absorbed. It left behind a sixty-kilometer plain of utter devastation, filled with thick mud, the detritus of houses and furniture, piled-up ice, rocks, and vegetation, and, inevitably, the corpses of men, women, and thousands of animals. Lulled by the apparent success of Venetz's tunnel, the early alarm system in the valley had failed, giving the remaining residents minutes rather than hours to escape. Most villagers had already relocated into the hills, however, and so the human tragedy was not on the scale it might have been. Venetz's brilliant, brave engineering on the dam had certainly saved Martigny, the largest town in the valley, from total destruction.

Nevertheless, the impacts of the Giétro debacle were devastating enough. A full decade after the event, an English travel writer named William Brockedon was struck by the wholesale "desolation and dulness" he encountered in the Val de Bagnes, especially as compared to the picturesque beauty of the neighboring valleys. Ascending the paths that crisscrossed the now becalmed River Dranse, he came across the ruins of a stone house, which stood like "an object of malediction," and symbolized "the desolate and ruined state of the valley." He itemized the geological changes wrought by the 1818 debacle:

> Vast blocks of stone, which were driven and deposited there by the force of that inundation, strew the valley, and sand and pebbles present an arid surface where rich pasturages were seen before the catastrophe. The quantity of the water suddenly discharged . . . and the velocity of its descent, is a measure of force which it is difficult to conceive.[20]

Another literally earth-shattering event, and another moonscape—courtesy of faraway Tambora. Ironically, despite this long-term devastation of the valley, Venetz's success in moderating the flood and ensuring a low death toll meant that his renown would remain local. The 1818 inundation of the Val de Bagnes occupies only a modest place in the rank of nineteenth-century European catastrophes, a relatively minor instance of the global tsunami of ecological consequences flowing from Tambora's 1815 eruption. As a turning point in the history of geology and climate science, however, the Giétro debacle assumes epochal proportions. The geological impact of the flood impressed Ignace Venetz as proof positive of far deeper historical processes in the Alps. These processes in turn demanded an entirely new kind of science—one founded on the concept and historical reality of climate change.

A CATASTROPHE, BUT NOT CATASTROPHISM

Now largely forgotten outside Switzerland, the 1818 Val de Bagnes disaster was nevertheless widely reported in the European press and became a major talking point in scientific circles. For those whose information came only from reading reports of the deluge, the event appeared to support the conventional catastrophist theories of geological formation, which, influenced by the biblical account, emphasized the shaping power of a great flood or floods that had once submerged the continent and carved out its valleys and mountains. This catastrophic diluvian scenario purported to explain the transport of erratic boulders far from their original location, as well as the thread of moraines at sometimes great distances from the current location of Alpine glaciers.

From a distance, the bursting of the River Dranse dam offered a very useful simulation of large-scale flooding, a kind of test case for catastrophism. Moreover, a selective sketch of the results proved highly reassuring to catastrophists. High above the valley floor, the Dranse flood had left new lines of debris that corresponded well with the character of ancient moraines. In addition, its tidal power had detached large boulders from the mountainsides and dumped them at great distances along the valley. One such block was measured at forty cubic meters, which, while still only one-tenth the size of the massive Pierre à Bot, the most celebrated of the Alpine erratics, seemed to confirm the transportive power of a massive torrent of water and mud, and to eliminate the need for any alternative geological explanation.[21]

Such, at least, was the general consensus surrounding the Val de Bagnes debacle. But the closest expert witness to the event, Ignace Venetz, was not convinced. His experience of the catastrophic flood of 1818 brought him, instead, to the diametrically opposite conclusion: only glaciers had the power to form the Alps. Two years earlier, he had delivered a paper that conformed to a traditional theory of erratic boulders transported by rolling on top of glaciers. By 1821, he had developed the outlines of modern glacial theory and periodic Ice Ages. In between, he met Jean-Pierre Perraudin and witnessed firsthand the catastrophic flood of the Val de Bagnes.

Venetz was a brilliantly intuitive geologist but, unfortunately, not a prolific or confident writer. His 1821 prize-winning paper to the Swiss Society is a rambling amateur affair, immersed in details, but its bullet-point conclusions sketch out, in bold terms, the basic principles of modern climate science. Glaciers were nature's own antique ruins, the "relics of former climates."[22] "The moraines found at a significant distance from the glaciers," Venetz writes, "date from a period lost in the mists of time." Therefore, by a simple but crucial step of logic, he must infer that "temperature rises and falls periodically, though in an irregular cycle."[23] Climate change, Venetz concludes, has driven an historical cycle of glaciation, which in turn has left its indelible mark on the geological formation of the Alps and by implication the European continent. Amazingly, Venetz's historic 1821 paper was not published for another twelve years, by which time Jean de Charpentier, after his own protracted period of doubtful rumination, had taken up the cause. After seeing to the publication of Venetz's paper, he promoted its conclusions in a far more widely read article of his own published in 1834, at which point the new glacial theory came to the attention of the new head of the Swiss Society, Charpentier's onetime protégé Louis Agassiz.[24]

We can wonder at how the course of nineteenth-century science might have been different had Venetz possessed skills of argument and self-promotion equal to his resourcefulness as an engineer and geological theorist. But rarely does such a combination of talents reside with one individual. More curious, therefore, and more profitable, is to speculate upon how it was that Venetz was converted to the radical glacialist theories of Jean-Pierre Perraudin in the aftermath of the Val de Bagnes deluge, when so many other observers saw in that event only a confirmation of the received wisdom—that a diluvian catastrophe had shaped the geological history of the Alps. On what grounds did Venetz come to the opposite conclusion and thus initiate the slow march toward the scientific truth of glaciation?

Where casual or more distant observers had seen proof of diluvian theory, Venetz had the benefit of close examination of the gorge and hillsides of the Val de Bagnes impacted by the violent deluge. There he found no new striations on the surface of rocks. While it was true the rushing tide of water had left moraine-like lines of debris at high elevation and had displaced rocks in large quantities, neither of these actions was on a scale to allow him to persist in the belief that water or mud alone could have been the agent of Alpine formation. With flooding eliminated, only a theory of glacial transport remained.

Though Venetz left no detailed account of the progress of his discoveries, there must have been a day, in 1819 or 1820, when his thoughts returned to his dramatic summer in the Val de Bagnes. There, amid all the pressures of the dam crisis on the River Dranse that it was his professional duty to resolve, his eccentric local guide had pestered him with wild ideas about mile-high glaciers. In the urgent anxiety of those days, he had paid them little mind. But now, as he cast his eye across the devastated valley with its unrecognizable, moon-like terrain, he must have realized that Jean-Pierre Perraudin was right. One couldn't rely on the evidence of one's senses or on mere common sense. The Earth was capable of radical and total transformation, a fact that required a great leap of imagination to accept. After the close-up, traumatic, life-changing experience of the Val de Bagnes debacle, Venetz was ready to make that leap.

It took a village to derive the modern theories of climate change and the Ice Age. Isolated speculations on glacial theory in prior decades by James Hutton and others had gone nowhere, lost in that strange histor-ical limbo of unrecognized truths. The chance meeting of Perraudin, Venetz, and Charpentier in the cold Tamboran spring of 1818 lit the initial fuse of speculation, while a second crucial meeting of glacialist minds occurred almost two decades later in the summer of 1836 when Louis Agassiz arrived with his family for a holiday at Charpentier's mountain chalet at Bex. Charpentier had been careful to invite Venetz to join them, and after several all-night disputations on Alpine geology, the three men decided on a professional bet on the merits of glacial the-ory. They abruptly left their families behind for a months-long scientific expedition across the Alps.

Agassiz, who thought Venetz's notions "bizarre," set out in the convic-tion that he would win the wager. He returned with the equally ardent conviction that he had been wrong. Venetz and Charpentier had con-verted the most influential scientist in Europe, for whom glacial theory would become an obsession for the remainder of his glittering career. Unlike either Venetz or Charpentier, Agassiz worked quickly and deci-sively, and had a talent for publicity. The very next year he delivered his famous address at Neuchâtel in which he surprised his audience of aca-demic worthies—who were expecting a satisfying discourse on fossils—with claims that the very place in which they were sitting had once been covered by a vast ocean of ice that stretched from the North Pole to the Mediterranean Sea. Like Perraudin and Venetz before him, it was now Agassiz's turn to be gazed upon with pity and irritation by respected men of science, as if he had just that minute gone stark raving mad.[25]

Percy Shelley's "Mont Blanc" has long been a favorite of the under-graduate classroom. Generations of English professors have presented it as a manifesto of Romantic thought in which the poet exalts the human mind as a god-like vessel of the world. Shelley's poem boasts one of the most celebrated openings in Romantic literature: "The everlasting universe of things/Flows through the mind, and rolls its rapid waves." History is full of coincidences in which "the everlasting universe of things" flowing through the human mind takes the same course through several minds at once. So it was with theories of glaciation in the Tambora period 1815–18. A Swiss engineer, a mine supervisor, and a chamois hunter each experienced his own version of Shelley's vision of a glaciated Europe. Together they embarked on a halting, intermittent collaboration that would evolve, two decades later, into a formal theory of climate change and cyclical Ice Ages, ideas that constitute the foundation stone of both modern geology and climate science.

Shelley's image of the mind's "rapid waves" of impression reappears later in "Mont Blanc" in his account of the Bossons glacier, where the same aquatic imagery turns abruptly sinister. In section 4 of the poem, the "flow" and "waves" of the creative imagination become the destructive actions of the glacier, before whose immense power human beings shrink "in dread." Here the advancing Bossons glacier is not an image of human imaginative power, let alone a picture-postcard vista, but rather "a city of death,"

> ... distinct with many a tower
> And wall impregnable of beaming ice,
> Yet not a city, but a flood of ruin
> Is there, that from the boundaries of the sky
> Rolls its perpetual stream; vast pines are strewing
> Its destined parth, or in the mangled soil
> Branchless and shattered stand; the rocks, drawn down
> From yon remotest waste, have overthrown
> The limits of the dead and living world,
> Never to be reclaimed. (105–14)

The remorseless glacier of Shelley's imagination destroys all plant and wildlife—"The dwelling-place/Of insects, beasts, and birds"—obliterating "life and joy" and all trace of human community: "The race/Of man, flies far in dread; his work and dwelling/Vanish." Shelley's great poem, at once a celebration of human creative powers, is haunted at the same time by the specter of universal glaciation, which would bring about an historical end to all human "works," including, inevitably, Shelley's own poetry. Hence the mood of gloom from which the poem never quite escapes.

Percy Shelley's "Mont Blanc"—like the atmosphere of icy doom in volume 2 of Mary Shelley's *Frankenstein*—prefigures the glacial disaster that took place two summers after their Alpine tour along the same chain of mountains. The catastrophic inundation of the Val de Bagnes in June 1818, which destroyed villages and farmland across one of the most picturesque valleys of the Alps, was a singular geoclimatological event, being the remote consequence of a volcanic eruption half a world away and three years in the past. In an uncanny way, the Val de Bagnes flood reenacted the destruction of the Sanggar peninsula on Sumbawa by Tambora's boiling pyroclastic flows in April 1815. In the Alpine case the agent of destruction was instead a "lava" of water, ice, and mud issuing from the unstable mountain gorge.

The link between Tambora and the Val de Bagnes was not, of course, understood at the time. Now, however, scientists are able to analyze such relationships through the prism of teleconnection: the complex causal relationships that knit apparently disparate climatic and geophysical events around the globe. In this chapter I have extended the physical principle of teleconnection to the world of ideas. In terms of its importance to the history of science, the 1818 Swiss debacle was no ordinary natural disaster. The drastic Tamboran cooling of 1815–18, by extending the range of the massive Giétro glacier and spawning a disastrous *jökulhlaup* in the Swiss Alps, imprinted the ghostly image of long-ago glaciation on the pioneering mind of Ignace Venetz. From this sketchy intuition evolved, by fits and starts, a founding truth of the modern earth sciences: climate change.

CHAPTER EIGHT

THE OTHER IRISH FAMINE

There are no walls can stop hunger.
—IRISH SAYING

It is important to remember that the misery of the Tambora period in Europe—years of famine, disease, and homelessness—was borne overwhelmingly by the poor, who left scant record of their sufferings. For the middle and upper classes—including the Shelleys and their circle—the social and economic upheaval of those years presented only minor inconveniences. By contrast with the illiterate underclass, these affluent Europeans left voluminous accounts of their lives and impressions, including great poems like "Mont Blanc." To look at only their documentary record, therefore, can leave one with the misleading idea that the Tambora years were not exceptional in the turbulent history of the early nineteenth century. We must scrutinize closely what they wrote for clues to the experience of the silent millions who suffered displacement, hunger, disease, and death in the eruption's wake. From the bubble of privilege within which the Shelleys and their peers composed their brilliant verse and letters, it is possible to catch gleams of this benighted other world through which they mostly passed oblivious.

The young London poet John Keats, for example—a peripheral but admired member of the Shelley Circle—set out on a walking tour of

Scotland and Ireland in the Tambora summer of 1818. In Scotland, he dedicated sonnets to Robert Burns and danced a reel with the local girls, but his experience in famine-stricken Ireland, on the roads around Belfast, left him disgusted and dismayed. "We had too much opportunity," he wrote to his brother Tom, "to see the worse than nakedness, the rags, the dirt and misery of the poor common Irish." In a passage that shows Keats struggling to adapt his abundant powers of lyric expression to scenes of grotesque poverty, he describes his surreal encounter with an old woman seated in an improvised sedan chair held aloft by two beggar children, as if in grotesque parody of aristocratic manners:

> The Duchess of Dunghill—it is no laughing matter tho—Imagine the worst dog kennel you ever saw placed upon two poles ... In such a wretched thing sat a squalid old Woman squat like an ape half starved from a scarcity of Buiscuit [*sic*] in its passage from Madagascar to the cape—with a pipe in her mouth and looking out with a round-eyed skinny lidded inanity—with a sort of horizontal idiotic movement of her head—squab and lean she sat and puff'd out the smoke while the two ragged tattered Girls carried her along—

Keats, like any young writer might, weighs up the literary possibilities of the scene: "What a thing," he wonders, "would be a history of her Life and sensations."[1] What a thing indeed—except, of course, he didn't write that history, and neither did anyone else. This is not simply from a deficit of sympathy—Keats, like other middle-class tourists in Ireland in those years, expresses "absolute despair" at what he encounters—rather, it speaks to the yawning social gulf that existed between the educated metropolitan classes of Europe and the poor rural masses in the early nineteenth century.

Much emphasis in recent historiography has been placed on the problematic European encounter with different races and nations around the world in the colonial period. This focus can mute our sense of the heterogeneity of the European order itself, and the extraordinary mutual alienation that existed between geographic regions *within* Europe, even before the mass industrialization of the cities for which

Marx developed his theory of class struggle. Even to a sensitive, liberal-minded city poet such as Keats, the poor Irish peasant appeared barely human, "like an ape half starved." Can it be any wonder then that the English rulers of mostly rural Ireland, with less than poetic souls, were able to justify to themselves their indifference to the deaths of tens of thousands of their Irish subjects during the Tambora emergency of 1816–18?

"A SEASON DREADFUL AND MELANCHOLY"

William Carleton, the most popular Irish writer of the first half of the nineteenth century, had already overcome his humble beginnings in the rural north to achieve literary celebrity in Dublin when, in 1847, he made a return pilgrimage to his birthplace in the Clogher Valley. There, among the mountain villages of County Tyrone, he found tragic signs of the Great Famine. Landlords had taken advantage of the penury of their smallholding farmers to launch wholesale evictions on their properties, literally casting their tenants out to the wind and weather. One once populous village near Carleton's childhood home, called Ballyscally, "was now a scene of perfect desolation. Out of seventy or eighty comfortable cottages, [the landlord] had not left one standing."[2]

Carleton's rage and despair over the fate of the inhabitants of Bally-scally, and the millions of others of his countrymen in the grip of the Great Famine, fired his literary imagination. Instead of writing a novel about the current crisis, however, he looked back to the period of famine and pestilence of which he had firsthand experience, the Tambora years of 1816–18, which he had passed as an itinerant witness to the suffering of the rural poor in Ulster. He dedicated his 1847 novel, *The Black Prophet: A Tale of Irish Famine*, to the British prime minister Lord John Russell because he

> knew that the approaching destitution and misery would require all possible sympathy from every available source; and he hoped ... that by placing before the eyes of those who had only *heard of* such inflictions, faithful and

unexaggerated pictures of all that the unhappy people suffer under them, he might, perchance, stir that sympathy into active and efficient benevolence.[3]

Carleton's plea fell on deaf ears. It is a matter of historical record that Lord Russell—encouraged by an influential laissez-faire ideologue in Treasury named Charles Trevelyan—allowed a million British subjects to perish on the doorstep of the most powerful and affluent empire on Earth in the years 1845–49. It was a providential corrective—Trevelyan not-so-secretly believed—to the imperial burden of Irish overpopulation and underdevelopment.

Residual outrage at this near-genocidal event is nourished by the vast Irish diaspora now spread across the world. For all the nostalgic attachment to the motherland and its history, however, the life and culture of "pre-famine" Ireland eludes recapture. In early nineteenth-century Ireland, most peasants were illiterate, spoke Irish exclusively, and left no records of their lives and sufferings. This includes the traumatic events of 1816–18, for which even official records are scant. *The Black Prophet*, then—written by an Irish son of the land who "crossed over" into the metropolitan, English-speaking world of the empire—stands alone as a literary monument to that doleful chapter of Irish history, written from the viewpoint of the peasantry themselves. In that novel, Carleton set himself a melancholy, monumental task: to record "all the final terrors of a people on the edge of extinction."[4]

Carleton's account of the 1816–18 tragedy begins, as all Tambora stories must, with the Frankenstein weather of the Year without a Summer. He witnessed the same skies over the British Isles that attracted the scientific interest of Luke Howard and awoke Turner and Constable to the sublime subject of sunsets and clouds:

> The sun, ere he sank among the dark western clouds, shot out . . . a light so angry, yet so ghastly, that it gave the whole earth a wild, alarming and spectral hue, like that seen in some feverish dream.

Carleton, writing three decades after the event, has awarded the red volcanic skies of 1815 the retrospective power of famine and fever, col-

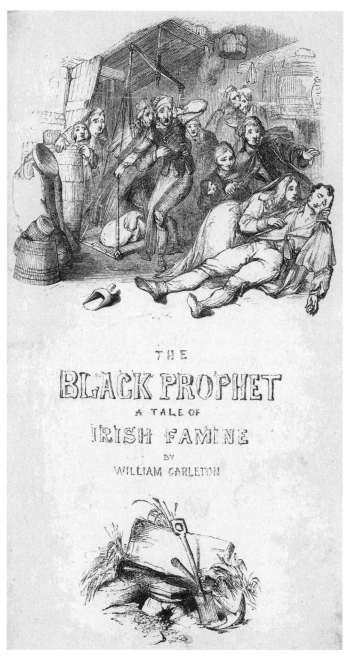

FIGURE 8.1. The title page from William Carleton's *The Black Prophet* (1847) shows the hero fainting from hunger, while beneath the author's name is the haunting image of a freshly dug grave.

lapsing them in his narrative memory with a very different canvas of sky—the relentless bitter cold and rain of 1816:

> The sky was obscured by a heavy canopy of low dull clouds that had about them none of the grandeur of storm, but lay overhead charged with those wintry deluges which we feel to be so unnatural and alarming in autumn, whose bounty and beauty they equally disfigure and destroy.... The whole summer had been sunless and wet—one, in fact, of ceaseless rain, which fell day after day, week after week, and month after month, until the sorrowful consciousness had arrived that *any* change for the better must now come *too late*, and that nothing was certain but the terrible union of famine, disease, and death which was to follow.[5]

Ireland's western, ocean-bound situation placed it in the vanguard of the brutal westerly storm systems tailing in from the Atlantic in the summer of 1816. The combination of weather deterioration and widespread preexisting poverty among the rural population meant a perfect storm of calamity for the Irish people. It being both wet *and* cold that summer conspired to kill their subsistence crops while also encouraging the spread of typhus-bearing lice, which attacked en masse their already weakened frames.

First, the weather report. During Ireland's "Year without a Summer," unwelcome Arctic ice lingered off the west coast, while 31 inches of rain fell over 142 days, mostly in the crop-growing months between May and October.[6] In Drogheda, ducks were reported swimming across the fields sown with oats and potatoes, while someone mailed a damp husk of green corn in protest to Robert Peel, the Chief Secretary for Ireland, at Dublin Castle. One doctor in the north of Ireland called 1816 a summer "to which I believe the memory of man furnished no parallel, being wet, cold, and in every respect incongenial to the growth or maturation of the fruits of the earth." The wheat crop failed, the grain small and blighted, or bursting its husk prior to germination. Bread made from the affected flour was inedible, so children took to rolling the damp lumps into balls and throwing them against the walls where

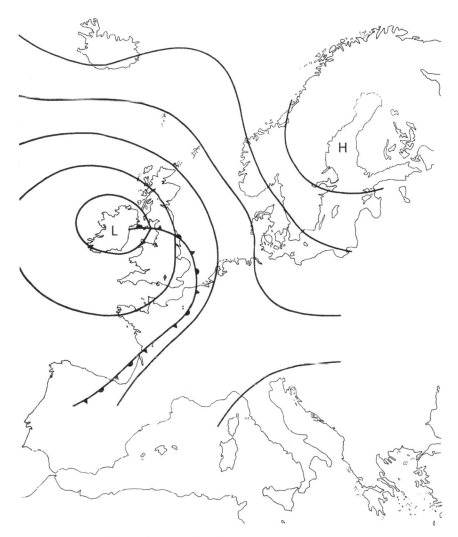

FIGURE 8.2. A synoptic weather map for July 7, 1816, based on reconstructions by pioneering historical climatologist Hubert Lamb. The map shows a storm-rich low-pressure system—remarkable for its unseasonality—centered directly over Ireland. (C. R. Harington, ed., *The Year without a Summer? World Climate in 1816* [Ottawa: Canadian Museum of Nature, 1992], 363; Courtesy of the Canadian Museum of Nature.)

they stuck like gum. Draught horses fell dead in their harnesses from the paltry nourishment of the season's oats.[7]

As we saw in the example of James Jameson's report on the Bengal cholera in chapter 4, members of the medical profession served as de facto meteorologists in the early nineteenth century. For the Tambora period in Dublin, the task of scientifically assessing the miserable weather was taken up by Dr. Francis Barker, who commented closely on the haywire dynamics of wind and rain:

> At this period the weather did not seem to depend on the direction of the wind so much as usual. In general, winds blowing from the northern points are, in this country, attended by dry weather; but during the summer and autumn of these years, from what quarter soever the wind came, it was accompanied by rain.[8]

The mean temperature in Dublin between February and October 1816 fell 3.5°F below average, while the rainfall in July, the heart of the growing season, was more than four times the amount of the corresponding month a year prior. With rain-saturated depressions churning above month after month, some parts of the island experienced double or more their average rainfall.[9] According to Barker, "the humidity of the atmosphere was almost incessant" through 1816, while in the opening pages of *The Black Prophet* William Carleton describes a sinister haze over the land that seemed to prefigure the end of the world:

> Long black masses of smoke trailed over the whole country, or hung, during the thick sweltering calms, in broad columns that gave to the face of nature an aspect strikingly dark and disastrous. . . . A brooding stillness, too, lay over all nature; cheerfulness had disappeared, even the groves and hedges were silent, for the very birds had ceased to sing.[10]

One characteristic of the biblical apocalypse is that all things are transformed into their opposites, rendering the familiar world an object of terrifying strangeness. Such was the emotional impact of the weather of 1816, during which "all those visible signs which prognosticate any particular description of weather, had altogether lost their significance."[11] For most of the world's population, dependent as they were on subsistence agriculture and the benign, predictable progress of the seasons, Tambora's bizarre weather must have induced a stomach-churning bewilderment and anxiety. The reactions of the Irish peasantry to their climate crisis traversed the spectrum from violent rage to drawn-out despair. In this, their moods mirrored the dark impulses of the skies overhead.

"THE TERRIBLE REALITIES OF 1817"

The unprecedented wet, stormy weather of 1816 continued into the following year. In February, newspapers reported "a storm, of singular awfulness, raged over the city of Dublin the whole of Thursday morning last, accompanied with loud peals of thunder, frequent and vivid lightnings, and the heaviest showers of hail and rain."[12] As conditions grew desperate through 1817, men walked tens of miles to buy cattle feed for the family table. When the money ran out, they sold their livestock, furniture, and finally their clothes. In a tragic irony, the peasants' rational preference for food at the expense of clothing worked against their survival. As their clothes turned to rags and blankets grew scarce, typhus-bearing lice were able to circulate freely within and between households, spreading disease at a breathtaking rate.

Poverty and intermittent famine had been a chronic issue in Ireland for centuries, but the scale of destitution in the 1815–18 period, owing to an increased population, came as a severe shock because the just-concluded Napoleonic Wars had been a time of relative prosperity. The late eighteenth century had witnessed a general upward trend in commodity prices, while the rapid industrial expansion of Britain created rising demand for Irish goods. Then, in wartime, when Britain faced significant restraints on her trade, Irish grain and linens fetched high prices. Standards of living improved, and the population continued its steady increase. But two entirely unconnected, epochal events of mid-1815—the eruption of Mount Tambora in April followed by the final defeat of Napoleon at Waterloo in June—combined to destroy Ireland's fragile economic growth. As one famine historian puts it, "the fall in agricultural prices after 1815 punctured the veneer of wealth and exposed the frailties beneath."[13]

Families that had enjoyed an affluence unknown to their own grandparents were suddenly cast into poverty in 1815 and 1816. That fine Sunday suit was quickly sold or worn every day until it hung like rags. Thus, when the bad weather came in the summer of 1816, it fatally amplified already severely depressed social conditions. "Seldom," reflected another eyewitness of the post-Tambora misery in Ireland, "had such a multiplication of evils come together."[14]

In *The Black Prophet*, William Carleton offers a vivid description of the impact of crop-destroying rains in 1816: "[They] took a short path across the fields, whilst at every step the water spurted up out of the spongy soil, so that they were soon wet nearly to the knees, so thoroughly saturated was the ground with the rain which had incessantly fallen."[15] From Kerry to Cork, and Donegal to Clare, the unfortunate farmers of 1816 witnessed the full gamut of rain-damaged crops, including waking to find their corn crop caked in red volcanic dust.[16] Most direful of all, the saturated soil created a toxic environment for the peasantry's subsistence food, the usually hardy potato. In damp ground, the watery film on a growing tuber's surface will restrict oxygenic diffusion. But once depleted of air, the cell membranes of the plant begin to collapse and leak, reducing resistance to infection, at which point a multitude of pathogens may stake their claim. Blackleg, soft rot, white mold, and powdery scab are just some of the picturesque names given to potato blight in overirrigated conditions. In short, the extended periods of heavy rain during the summer of 1816 first exposed the fallibility of the Irish potato crop. Tragically, few measures were subsequently taken to reduce Irish reliance on potatoes, with calamitous consequences in the 1840s.

Ironically, the potato was widely considered a breakthrough subsistence crop in Europe because it was less vulnerable to meteorological variability. What this view did not take into account, however, was climate change—when the potato crop faced a sustained period of extreme weather events outside the range of natural variability. In good years, the rural population subsisted on meals of potato, buttermilk, and oatcakes, but in 1816 and early 1817 even these subsistence foods became scarce and expensive. Starving people roamed the woods in search of "ramps"—a wild onion considered disgusting in ordinary times. Girls shaved their heads and sold their hair to peddlers for a pittance, while families bled their half-starved cattle, feeding on the blood mixed with a little barley—truly a soup of the damned.[17]

In 1816, the soaked earth also ruined the quality of peat soil on which the peasantry relied for heating their cabins. No dry straw to sleep on—instead just the damp earth. It was thus in the first winter after Tambora that the disease ecology specific to typhus began to emerge from the deteriorated living conditions of the Irish poor. Clothes turned filthy from overwear, while threadbare coats and blankets were shared among the family. As the cold weather settled in and with no fuel to burn, families huddled together in their cabins for warmth, often not venturing out of their beds for days at a time. One doctor in County Tyrone reported to have "frequently found all the members of the family laid in the same bed with a patient labouring under fever, owing to their having but one or two blankets."[18] Even as famine conditions subsided with the improved harvests of 1817, the Irish peasantry faced the even greater horror of epidemic disease. In their already weakened condition, they had few resources to fight it.

ONE LOUSE, TWO LICE

As any parent of schoolchildren knows, the louse is always with us. Indeed, the shared history of humans and lice constitutes a remarkable instance of co-evolution.[19] Five million years ago, when the ancestors of modern human beings diverged from the chimpanzee, a new species of lice (*pediculus humanus*) joined us on our evolutionary adventure. A second historically monumental divergence occurred when human beings began to wear clothing. Ambitious lice migrated to the human body (*pediculus humanus humanus*), where they developed the nifty expedient of depositing their eggs in human clothes and taking up residence there. Striking evidence of our close, co-evolutionary relation with the louse is its slim genetic signature, characterized by a deficit of genes associated with environmental sensing and precious little metabolic engineering. The domesticated louse has no need for receptors of smell or taste, and possesses the smallest number of detoxification enzymes of any insect. Why wander about, when the human body and its vestments offer the coziest possible accomodation?[20]

It is unclear exactly how and when the modern typhus bacterium (*rickettsia prowazekii*) emerged to fatally complicate the host-parasite relation between humans and lice. Some argue that the famous plague of Athens described by the historian Thucydides was in fact typhus. More likely, however, is that typhus resulted from the chaotic biological exchange initiated by European colonization of the Americas in the fifteenth century. One conventional hypothesis traces the typhus pathogen from the Far East to Spain in the fifteenth century and from there to the Americas and a global imprint. But the recent discovery of a nonhuman reservoir for typhus in the American flying squirrel suggests an inverted etiology, namely that—like yellow fever, cholera, and syphilis—typhus originated as a colonial disease brought *back* to Europe from the New World. The deadly typhus, according to this scenario, "was born in the chance meeting of an American rickettsia and a Spanish louse."

The impact on modern human history of this freak biological union is incalculable. Typhus decimated Napoleon's army on his disastrous retreat from Moscow. A century later, during the revolutionary period 1917–25, twenty-five million Russians contracted typhus, killing an estimated three million. More recently, in the 1990s, typhus reasserted itself as a major global threat, infecting over one hundred thousand people during the civil war in Burundi. In terms of sheer numbers of victims, louse-borne typhus has not only been decisive in the outcome of major military conflicts during the last five hundred years, it has in that time "killed more people than all conflicts combined."[21] The typhus epidemic that swept across peacetime Europe in 1817–18 is thus but one episode in a five-hundred-year biological disaster: an example of the vulnerability of human communities to modern, globalized pathogens in times of material distress, be it war, famine, or an ecological breakdown precipitated by climate change.

In one sense, the louse has gotten a bad rap in all this. After all, it is we, as the natural reservoir of the typhus rickettsia, who first infect *it*. The louse ingests the disease bacteria from the blood of its human host, the rickettsia carrier, whereupon it is mortally infected. Between infection and death, however, lies the window of opportunity for the louse to spread the contagion. If the host falls ill, the louse—highly sensitive to febrile body temperature—will seek out a new host. In prosperous times, where sanitary conditions prevail, this search will be fruitless. But if the louse is fortunate enough to find itself in a humid environment among a human community in crisis—housebound families unable to wash or change their clothes, sharing what coats they have, and huddling together under blankets for warmth—it will successfully migrate to an alternate body. Its bite is not its death warrant; rather, its dry and powdery feces, laden with typhus bacteria, infect the bite wound. As the bacteria multiply in the bloodstream, massive cellular damage affects the vital organs and gastrointestinal tract. Internal hemorrhages ensue. The victim, confused and feverish throughout an ordeal that may last as long as two weeks, endures painful edematic swelling, organ failure, and ultimately death in at least one in four cases. From there, it is only a matter of time and statistics. If the weather remains bad and the living conditions of the human community do not improve, the typhus is essentially unstoppable.

William Carleton recalled the fatal evolution of the 1816 famine into "universal" epidemic disease the following year in the Clogher Valley:

> the gloom that darkened the face of the country had become awful. . . . Typhus fever had now set in, and was filling the land with fearful and unexampled desolation. Famine, in all cases the source and origin of contagion, had done, and was still doing, its work.[22]

When typhus struck, the resources of rural Irish communities, already hard-pressed by critical food shortages, were stretched to the breaking

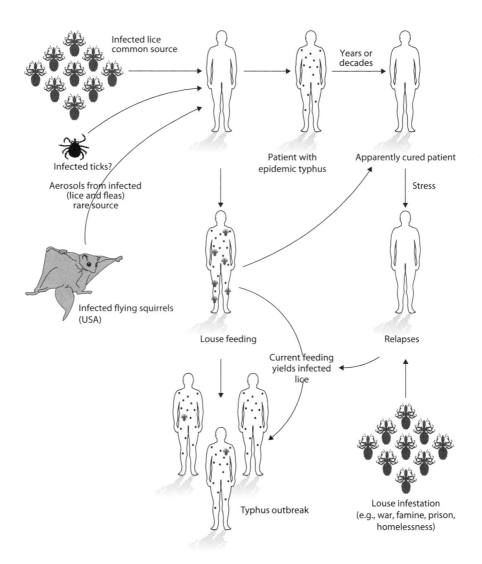

FIGURE 8.3. The vectors and life cycle of the typhus bacterium *rickettsia prowazekii*. Note that a person infected with typhus may carry the disease asymptomatically for years, before triggering a new outbreak under the stress of deteriorating social conditions. (Yassina Bechah et al., "Epidemic Typhus," *Lancet Infectious Diseases* 8 [2008]: 420.)

point. The English travel writer John Trotter paints a pitiful picture of villages, not necessarily remote, that found themselves beyond the reach of charity or government assistance in 1817, simply abandoned to their fate:

> Poor mud cottages were scattered along the road-sides, and we learned, with heartfelt sorrow, that fever was spreading everywhere among them. When this infectious malady enters his cottage, the Irish peasant and family are the most wretched of human beings! Unable to procure medical, or any other aid,—provided with no matters useful for the sick,—and becoming objects of terror in the midst of their poor and uninfected neighbours,— they sicken, linger, and die in their habitations.

Everywhere he went, walking along deserted streets, Trotter could hear the groans of people dying abandoned and alone, or entire families perishing together once the last nurse among them had fallen ill. "The heart sickens," he wrote, "under the repeated observation of so great a mass of human wretchedness, and our toilsome way has been frequently made insupportably painful to us by it."[23] Trotter wondered, as we do in reading his account, where was the help for these pitiful victims, dying in their tens of thousands across the Irish countryside?

Unfortunately for the suffering rural population of Ireland, no effective system of relief existed. The numbers of English absentee landlords, who drew profits from their Irish estates only to spend the money in London and elsewhere, meant a fatal breach of the feudal contract: the gentry weren't present to witness the suffering of their tenants and so felt minimally motivated to help them. Nor had the government, which complained loudly about absenteeism, moved to supply the deficiency. In Cork, the liberal reformer William Parker lamented at the outset of the famine in 1816 of "nearly a total absence of all legislative provision for our Poor." As climate refugees flocked into the city, Parker called the "mass of distress" on the streets of Cork "a disgrace to a nation professing the principles of the Christian Religion."[24]

Such appeals were to no avail. In February 1818, after almost three years of rapid economic decline culminating in a full-blown famine and epidemic, a group of desperate Cork citizens petitioned the British Parliament "to protect them from that ruin into which all ranks appear to be fast sinking."[25] The petition was tabled without comment. Less than a year before, the same Parliament had involved itself in lengthy debate on the prospects for Catholic Emancipation. That no corresponding attention was paid to the very present disaster of Irish poverty and disease shows the tragic consequences of a nineteenth-century Anglo-Irish political discourse dominated by sectarian arguments. While Parliament wrangled over Irish souls, across the Irish Sea the body count was steadily mounting.

John Gamble, a minor society writer from County Tyrone, living in London, never enjoyed the literary fame of William Carleton, his fellow Ulsterman. But on a return visit to his hometown of Strabane in the summer of 1818, he recorded unforgettable impressions of Ulster at the tail end of the Tambora disaster: "Since I was last here, this town and neighbourhood have been visited by two almost of the heaviest calamities which can befall human beings. Fever and famine have been let loose, and it is hard to say which has destroyed the most."[26] Gamble's reaction echoes that of the doctor William Harty, who enumerated how the 1817–18 typhus epidemic impacted Irish society far beyond the raw numbers of infected and dead:

> The loss to society from the interruption given to productive labour; the expense incurred by providing for the sick; the debility and weakness of constitution induced by the disease; the mortality which must attend it, and is most frequent where it is most injurious, namely, among men advanced in life, who are often the heads and support of families; the increase of poverty and mendicity, together with the agonizing mental distress to which it must give rise, are consequences of this epidemic that must occur to every humane and reflecting mind.[27]

Harty's point: survivors of the epidemic faced a brutally diminished existence marked by poverty, dislocation, and family breakdown.

Back in the devastated town of Strabane, the "humane and reflecting mind" of John Gamble saw few familiar faces and received no joyful

welcome. Heartbroken, he looked across the empty streets and contemplated the massive depopulation that had occurred in his home county.
Some had emigrated, while many more had died of starvation and disease. Where a few years earlier he had seen a bustling high street full of
shops doing brisk wartime trade, now all was eerily quiet. "I walk therefore," he wrote, "nearly as much alone as I should in the wilds of America."[28] During the previous two years, one-and-a-half million Irish had
been infected with typhus, with probably in excess of 65,000 deaths. Up
to 80,000 had perished the previous year, 1816–17, from the famine that
preceded the epidemic. An unknown number, too, had fled the country
as the last best means of escaping the deadly pincer grip of these tragedies. Perhaps expecting a hero's return to his old town, Gamble instead
stumbled into the desolated landscape of the new Ireland—a famished,
traumatized Ireland—the Ireland of the nineteenth century.

"THE AMIABLE PECULIARITY OF THE IRISH CHARACTER"

The fact that there are almost no official statistics on the Irish famine of 1816–17, while the typhus epidemic of the subsequent two years
produced a veritable library of reports and treatises, signifies that the
ordinary suffering of the Irish peasantry mattered little to their urban
compatriots, and even less to their British rulers. But once the mal-
nourished masses took to the roads, bringing contagion with them, the
privileged metropolitan classes and their agents in the government and
media rose up in alarm. Parliamentary debates characterized the typhus
epidemic of 1817–18 as a security threat rather than a public health crisis,
thereby ignoring the systemic issue of Irish poverty and famine's close
relation to disease.

Along the length and breadth of Ireland in 1817, typhus fever had
spread "to an extent unprecedented in the recollection of any person
living."[29] And as the epidemic reached the cities, so did its refugees.
Exact numbers are unknown, but certainly hundreds of thousands
of Irish abandoned their homes and took to the open roads in 1816–
18, a demographic upheaval that fully exposed the fragility and stark

inequities of British Ireland under the Act of Union. A countryman of William Carleton's reported that during the first wave of the refugee crisis, "many hundred families, holding small farms in the mountains of Tyrone, have been obliged to abandon their dwellings in the spring of 1817, and betake themselves to begging as the only resource left to preserve their lives."[30] Reports from Limerick were that "the whole country is in motion," while in Derry, "almost the entire population" of the rural districts had left their homes for the towns.[31]

The emergence of mass refugeeism in almost every region in Ireland spurred a vigorous government reaction. Everywhere, city authorities floundered. Tullamore officials posted soldiers on all routes into the town to turn away the indigent crowds while, back in the imperial metropole of London, Parliament acted decisively. The "Act to Establish Regulations for Preventing Contagious Diseases in Ireland" is a legislative document of class panic and spectacular inhumanity, endowing health officials with powers "to apprehend all idle poor persons, men, women, or children, and all persons who may be found begging or seeking relief . . . and to direct and cause all such idle persons, beggars, and vagabonds to be removed and conveyed out of and from such parish and place." There is no word in the act on where these "idle" infected masses might then go, or be provided for. The peculiar indifference of the government to the fate of the Irish poor themselves did not pass unnoticed. Advocates for Ireland suspected the influence of genocidal ideology in the British government's legislative response to the crisis, bitterly denouncing "those pseudo-philanthropists, who can contemplate, not only without pain but with complacency, pestilence thinning the ranks of our 'superabundant' population; or who, to use the philosophic phraseology of our Malthite disciples, can, with unalloyed satisfaction, behold fever 'doing its business.' "[32]

While the poor in Ireland had at least a handful of advocates in Parliament, however ineffectual, they had no voice at all in the media. The metropolitan newspapers in Ireland, taking their cue from the government, systematically misrepresented the crises of 1816–18. First, the rural famine went largely unreported for fear of prompting speculation in the grain markets. Then, in early 1817, as typhus ravaged rural Ireland, newspapers in urban centers such as Dublin and Cork labeled the rumors of epidemic as "alarmism," shilling for a metropolitan merchant class concerned about the disruption of trade and possible quarantines. Even in the autumn of 1817, when typhus had reached the cities and its presence could no longer be denied, little sympathy was spared for the rural victims of famine and disease. Instead, the city papers demonized the starving refugees, calling upon authorities to bar the city gates against the typhus-bearing hordes. In Dublin—the metropolitan hub of Irish government and public opinion—editors railed against those "vagrants and beggars, who have emerged from receptacles of disease, and spread themselves in various directions." They demanded the government crack down on the beggary "which prevails to such a disgusting, and in the light we are now considering it, dangerous degree." In the grip of their paranoia of contagion, the newly arrived beggar, their countryman, was worse than a leper: "his touch, to whomsoever given—nay, even the very air which is about him, are pestiferous."[33]

On the outskirts of blockaded Irish towns in 1817, "fever huts" cropped up everywhere—"wretched structures of mud or stone" with straw roofs hastily erected along roadsides and in fields for fever victims with nowhere else to go. There the refugees "struggled with a formidable disease on the damp ground, with little covering but the miserable clothing worn by day, and scarcely protected from the inclemency of the weather." For the homeless, there was competition even for these pitiful dwellings. According to one account from Kanturk near Cork, a refugee family found a fever hut too small to accommodate their number, whereupon two daughters were forced to live outside on the open ground. When the father succumbed to typhus, his daughters then fought to take his place in the hut, the stronger one winning out against her sibling.[34] This book includes many stark vignettes of human wretchedness during the Tambora period, but surely none more obscene than this. More important is the general truth it elicits. To quote an Irish doctor of the time, "nothing short of extreme misery could have wrought so sudden and complete a change in the feelings of a people, whose attachment to their offspring and relatives is proverbial."[35]

Into the first week of September 1817, newspapers in the southern towns continued to hold the line against panic. "There is no ground whatever for alarm," the *Kilkenny Moderator* reassured its readers, while the *Sligo Journal* quoted the opinion of a "professional gentleman" who firmly believed that "the symptoms and operation of the existing fever are of a very mild description, and merely such as usually occur at this period of the year." In Dublin on September 11, the *Evening Post* considered itself "authorized to declare it as their opinion, *that the Epidemic Fever of the Country does not pervade this City!*" Even from their tone, it is clear the editors doth protest too much. For by this time, the newspapers in the north of the country were reporting "unprecedented numbers now dying of Fever." Typhus was "raging" in Enniskillen, while from Strabane, in County Tyrone, came reports of a shortage of carpenters for building coffins. By mid-September the game was up, and subscribers to the *Dublin Evening Post* were called upon to digest the following solemn announcement: "We regret to learn, that the Fever, in the vicinity of Dublin, has assumed a very *malignant type*." Two months later, the editors had abandoned all rhetorical restraint: they reported the typhus "deepening and spreading with the rapidity and ravages of a Plague."[36]

Amartya Sen, the Nobel Prize–winning economic theorist of famine, has identified "the nature and freedom of the news media" as critical to whether food shortages will escalate into general famine. If information about localized problems flows freely, and media pressure is applied to governments to act, many incipient famines may be prevented.[37] No such responsible fourth estate existed in the Ireland of the late 1810s. But if the journalists of Ireland emerged with little credit from the disaster—whipsawing from denial to doom-filled pronouncements— the same cannot be said of the clergy and medical men, many of whom risked their lives to organize relief and personally care for the sick. Thirteen priests died attending typhus victims in Carleton's hometown of Clogher alone, while for the medical profession, still in its infancy, the Tambora crisis was their finest hour.[38] As rising members of the civic establishment, prominent Dublin doctors such as Francis Barker (the amateur meteorologist) and William Harty urged new public health initiatives upon the recalcitrant Irish elite and their English colonial masters.

As part of their improvised insurgent strategy, the doctors attached themselves to the new Association for the Suppression of Mendicity (street begging) in Dublin. As self-appointed members of the "subcommittee of health" in early 1818, they began to agitate for an enlarged public health system and anti-poverty measures to be introduced to address the root causes of both beggary and disease. Harty and Barker's progressive agenda did not meet with a ready welcome in the offices of the Lord Lieutenant of Ireland, however, who distanced himself from the association and bluntly refused to provide government funds to alleviate the crisis. The doctors' plans, he averred, "can be more effectually carried into execution by private exertions and parochial subscriptions."[39] In other words, the usual organs of charity and the church should bear the burden of humanitarian relief, as they had under the Old Regime.

The doctors were undeterred, however. Through the desperate weeks of early 1818 the health committee continued to publish embarrassing resolutions calling the government response "totally insufficient" to the magnitude of the ongoing crisis. Decades ahead of their time, these Dublin doctors preached "a preventive system . . . calculated to avert an immense accumulation of wretchedness and poverty." They went so far as to call the Lord Lieutenant's expressed reliance on private charities a "fatal delusion"—strong language that, in 1818, retained more than a whiff of Jacobin revolutionary spirit about it. They expressed outrage at the rejection of their plan to mandate the cleaning of houses infected with fever and bitterly denounced the efforts of Dublin authorities to downplay the extent of the epidemic for fear of its impact on trade. "Can we," they railed, "with such examples before our eyes, vilify Mahometans, and abuse their stupid indifference (under better motives) to the desolating devastations of the Plague?"[40]

The coup de grâce of the militant doctors' campaign came in September 1818, when the Association for the Suppression of Mendicity evicted the vast number of beggars it had itself accommodated and let them loose upon the affluent neighborhoods of Dublin. Their goal?

To shame the government and wealthy citizens of the city into coughing up relief donations. In what was for that time a remarkable public demonstration of class inequity and resentment, two thousand ragged beggars—many of them emaciated and sick—marched from the association's headquarters on Hawkins Street through the leafy squares of Dublin's elite, stopping to yell abuse at the houses of those known to have refused charitable aid. The impact of this daring piece of political theater was dramatic: almost ten thousand pounds in private donations, a huge sum, flowed into the association's coffers in the following days.[41]

At the height of the typhus epidemic, in September 1817, the Dublin doctors could loudly warn of "the ruin that awaits us, if every heart and hand are not speedily roused to exertion."[42] Even as late as October 1818, with typhus spread across the British Isles, the medical establishment in Edinburgh expressed alarm at "the state of continued fever . . . in our time we have never known it extend so generally over the Empire, or continue so long."[43] By the end of 1818, however, the worst of the epidemic was over, and with it the doctors' platform for agitating for public health reform.

Almost a year after the beggar's march in Dublin, Charles Grant, the new Chief Secretary for Ireland, felt comfortable in rising to his feet in Parliament to give the official government account of what had happened in Ireland over the preceding three years. Like Francis Barker and, later, William Carleton, Grant opened his narrative of the disaster with a meteorological description:

> In the years 1816 and 1817, the state of the weather was so moist and wet, that the lower orders in Ireland were almost deprived of fuel wherewith to dry themselves, and of food whereon to subsist. They were obliged to feed on esculent plants such as mustard-seed, nettles, potato-tops and potato-stalks—a diet which brought on a debility of body and encouraged the disease more than anything else could have done.[44]

Where Barker's interests in the weather were scientific, however, Grant's were political. The bad weather was to be one of a suite of causes Grant would offer for the death of some 150,000 of his Irish subjects from famine and disease in lieu of any acceptance of government responsibility for their deaths.[45] These included a particularly arch tactic of sugarcoating the blame for the spread of the typhus epidemic, then directing that blame toward the victims themselves. Native hospitality, "that amiable peculiarit[y] of the Irish character," had meant the sick were not quarantined as they ought, while the ritual waking of the dead exposed still more to the contagion. Also, the poor had failed to adequately "fumigat[e] their houses."[46]

Unfortunately for the historical reputation of the British Parliament, these explanations were either patently false or irrelevant. In many cases, the Irish locked their doors against the sick and indigent, an attitude abundantly evident in the mass evictions of rural cottiers and the anti-beggary editorials of the Dublin press. To offer an example: Thomas Mellon, who was to become patriarch of one of the richest banking families in the United States, spent his boyhood on a moderately affluent farm in County Tyrone. Late in life, he remembered that when his parents left him to go to the market during the Tambora years, they left him strict instructions to bolt the farmhouse door and a fierce dog to help repel the "tramps" that were "numerous at that time."[47]

Perhaps aware of the disingenuous nature of his tribute to Irish compassion, Grant moved quickly in his speech toward safer rhetorical ground: a glowing panegyric on Irish stoicism. The "patience" with which the Irish had borne their suffering was "truly admirable," he declared to an approving parliament; in fact, it "was not to be paralleled by anything in history." Despite their acute miseries, the population had not been moved to "the slightest tumult" or riot, while the "general benevolence" of the clergy and medical profession was "beyond all praise." Only at the very end of his lengthy speech did Grant allude to the minimal actions of his own government during the crisis by restating the unalterable principle of laissez-faire economics that no government should interfere with the workings of the marketplace. Although a public works program might be temporarily introduced to relieve unemployment, "any permanent legislative enactment on such a subject would be nothing more than a delusion." With that, Charles Grant sat down, "amidst considerable cheering from both sides of the House."[48]

While allowance should be made for the harsh ideology of Grant's speech—laissez-faire notions of "political economy" stood unchallenged at the time—little excuse can be found for his premeditated falsehoods. The most egregious was his claim that there were no civic disturbances in Ireland during the height of the crisis of 1817–18. Grant would have known this to be untrue, and not simply from his privileged access to government documents and deliberations. As early as May 1817 newspapers had carried reports of riots in Kildare and of an entire county "bordering on rebellion."[49] Starving mobs looted the granaries and attacked food convoys on their way to Dublin for export. Similar violent incidents were reported in Cork, Limerick, and Waterford. On March 4, 1817, the *Times* of London reported a full-blown "rising" in Tullamore, where "carts and cars have been broken, potatoes and meal seized and forcibly sold; the luggage and provision-boats stopped on the canal; and menaces thrown out that the locks would be smashed and the banks thrown in if the provision-merchants attempted to convey the produce of the county to Dublin." In Ballina, the army was called in to break up a riot over the export of oatmeal. The soldiers opened fire, killing three of the protestors and wounding many more.[50]

For William Carleton, the train of disasters that befell rural Ireland in 1816–18 followed an inexorable course from extreme weather to crop failure, to famine, to epidemic disease, to violent civic breakdown: "When a nation is reduced to such a state, no eye but the eye of God himself can see the appalling wretchedness to which a year of disease and scarcity strikes down the poor and working classes."[51] As history further records, the Tambora crisis in Ireland marked the end of a period of relative prosperity in that country, and the beginning of an era of intermittent food shortages and political instability culminating in the calamitous crop failures, mortality, and epic social disintegration of the Great Famine.

Both in popular memory and in scholarly histories, the famine of 1845–49 stands as zero hour of modern Irish history, when the country faced a massive and traumatic depopulation from which it has never fully recovered. A million died in those years of the potato blight, while another million emigrated. This history, as indelible as it is, has tended to cast events prior to the 1840s disaster into the foggy netherland of "pre-famine Ireland." As the Tambora-era record shows, however, a disaster of comparable dimensions, if not length, struck many regions of Ireland in 1816–18 when, for the first time, the subsistence potato crop failed across the country. In the large-scale famine, social breakdown, and epidemic conditions that ensued, the Tambora period offers a nightmarish prequel—a "black prophecy"—of the calamity that would shatter and transform Ireland a short generation later. Carleton, with the novelist's power of metaphor, offers a graphic image of the deep symbiosis between climate change and human destiny in the dystopic Tambora period:

> The very skies of heaven were hung with the black drapery of the grave; for never since, nor within the memory of man before it, did the clouds present shapes of such gloomy and funereal import. Hearses, coffins, long funeral processions, and all the dark emblems of mortality, were reflected, as it were, on the sky, from the terrible works of pestilence and famine, which were going forward on the earth beneath it.[52]

What are the lessons of the "forgotten" Irish famine of 1816–18? First of all, weather deterioration provides only the initial conditions for a humanitarian disaster. Much more depends in the longer term on the resilience of the communities affected, on their flexible will and capacity to adapt to drastic environmental changes, and on the resources of government. The nation-states of Europe—and particularly Britain in its responsibilities for Ireland—largely failed this test in Tambora's aftermath, and were rescued only by the return of seasonable weather in mid-1818 and the subsequent bountiful harvests.

In Doctor Frankenstein's ambivalent feelings toward his humanoid creation, we can trace the same dehumanizing impulse that allowed many among the metropolitan affluent classes of Europe to abandon legions of the rural poor to their miserable fate in 1816–18:

FIGURE 8.4. This remarkable hybrid illustration from Carleton's *The Black Prophet* (p. 27) shows the "sky" above the lovers' heads filled with phantasmic scenes of human suffering from the Irish famine and epidemic of 1816–18. Following Carleton's text closely, the illustrator represents the calamity as meteorological in origin, where rainclouds shape a nightmarish vision of fever-stricken victims, deathbeds, and funeral wakes. The contrast with the sentimental image of the hero and heroine is jarring.

I compassionated him, and sometimes felt a wish to console him; but when I looked upon him, when I saw the filthy mass that moved and talked, my heart sickened, and my feelings were altered to those of horror and hatred.

A eugenic loathing characterizes all of Frankenstein's interactions with the Creature. From that visceral "hatred" engendered by his re-pulsive appearance, Frankenstein arrives at the same conclusions on the monster's proper fate as the British government did with regard to Ireland's "superabundant" population in the early nineteenth cen-tury: they should be encouraged to emigrate if possible, and they should *not* be allowed to reproduce. Accordingly, the Creature offers to leave Europe for South America, while Frankenstein later destroys his half-made bride. "Thou didst seek my extinction," the Creature cries in his parting words to his dead creator, "that I might not cause greater wretchedness."[53]

That said, the class politics produced by the Tambora emergency were not all reactionary and inhumane. Both Robert Peel and Charles Grant, for all their foot dragging and parsimony in terms of humani-tarian relief, were moved by the Irish famine and epidemic of 1816–18 to begin conceiving a modern public health bureaucracy to cope with emergencies on a national scale. Peel empaneled a national fever com-mittee that evolved under Grant to become the first Board of Health in the British dominions. In 1817, the British Parliament in turn passed the landmark Poor Employment Act, which authorized a process for public loans to fund privately managed infrastructural projects to alle-viate unemployment. After 1817, public works programs became a stan-dard feature of economic policy.[54] Nineteenth-century Ireland, in ways good and bad, served as Britain's social laboratory. "It is no exaggeration to say," concludes one historian, "that the welfare state in England was foreshadowed by events in Ireland in this period."[55]

The same generally progressive trend occurred on the Continent. The French and Prussian governments spent massively on grain imports in 1816–17 and intervened in the marketplace by selling their own food re-serves at throwaway prices. In northern Germany, private co-operatives, led by the affluent elite, bypassed sluggish authorities to import grain directly from Russia to feed the starving in their communities. Vitally, for the subsequent long-term stability of Europe in the nineteenth century, the sheer scale of the humanitarian disaster of 1816–18 initiated the reeducation of the ruling classes in their moral responsibilities to the broader citizenry. In the process, it weakened the grip of the extreme laissez-faire ideology that had characterized the first phase of European industrial modernization. Patrick Webb of Tufts University, an authority on nutrition and food security, has pointed to 1817 as a watershed year in the evolution of modern humanitarian theories of governance. Desperate relief measures adopted during the Tambora emergency "contributed to a growing public acceptance of government action in times of crisis, while establishing a variety of viable approaches that continue to be used today."[56]

Out of the global tragedy of Tambora, then, emerged the rudiments of the modern liberal state? It would not do to overestimate the pace of progress. In many cases, new welfare laws were not enforced, and the evolving humanitarian rhetoric of the nineteenth century remained just that. Moreover, no progressive argument whatever can be made for Britain's colonial dominions in Asia and Africa, in which racism and Malthusian ideology combined to inflate the human toll of climate-related disasters throughout the Victorian period.[57] Even close to home, typhus continued to afflict Ireland long after it had been eradicated from the rest of Europe. One thing is certain: the haphazard evolution of humanitarian ideals in the nineteenth century did not advance quickly enough for the twice-damned children of the Tambora emergency in Ireland, who survived that trauma only to perish in their hundreds of thousands in the Great Famine of the 1840s.

CHAPTER NINE

HARD TIMES AT MONTICELLO

From Indonesia to India, from China to the Alps, from the Arctic wastes to the villages of Ireland, our Tambora story has contained multitudes. We have sailed hemispheres and crisscrossed domains of earth, sea, and sky. Now finally we turn to North America, where the folk memory of the Year without a Summer has, arguably, endured longer than anywhere else. Writing in 1924, meteorological historian Willis Milham could nominate the disastrous growing season of 1816 as the most "famous . . . written about" weather event in American history: "If all the statements in climatologies, in books on the weather, in biographies, in histories, and in the periodical literature were collected, they would form a sizable volume."[1] Even at the the end of the twentieth century, 1816 continued "to be a topic of great popular interest," particularly in the newspapers and journals of New England.[2] Fascination with the lost summer of 1816 has, for two hundred years, been shared between meteorologists and popular historians, with a shelf-load of commentary to show for it.

The conspicuous gap that remains—and which this chapter aims to fill—is to rewrite the fabled Year without a Summer as a *nationwide* teleconnected narrative of weather disasters, demographic upheaval, and economic boom-and-bust that helped shape a full decade of the social history of antebellum America. The Tamboran deep freeze also signaled an end of the early republican era of strident climate optimism, embodied in the patriotic figure of Thomas Jefferson. Approaching the bicentenary of Tambora's world-altering eruption, it is time to rescue the Year without a Summer from the dusty back pages of American folklore: to reimagine the late 1810s in the United States as a multiyear period of extreme weather with cascading social and political effects— and hence marked relevance to the twenty-first century. In this chapter, the old weather legends revive again to haunt us, this time as premonitory images of our own emerging climate dystopia.

EIGHTEEN-HUNDRED-AND-FROZE-TO-DEATH

Residents of the isolated community of Annsville, New York, placed a high value on formal education, a scarce resource in rural America in the early nineteenth century. A schoolteacher's arrival was like "an angel's visit," a rare and precious event.[3] In the summer term of 1816, Annsville had been blessed by just such a visit, and the children of the scattered settlement in Oneida County dutifully set out each morning for the long walk to the schoolhouse. The unusually bitter, frosty mornings of early June did not dull the enthusiasm of two outlying Annsville households for educating their offspring. Nor did the fact their four children had one pair of shoes between them for the three-mile walk, and no stockings. What happened to this unlucky group of schoolchildren in the Year without a Summer became the stuff of fireside legend for generations of upstate New Yorkers.

On the morning of June 6, the four classmates, aged six to nine, set out as usual and arrived punctually at the schoolhouse. For a six-year-old to walk three miles barefoot in a frost must have been an ordeal unto itself. But worse lay in store. This strange, cold June day grew progressively *colder*, contrary to all the norms of summer in the Northeast. The temperature in the schoolhouse—none too comfortable at the best of times—had become intolerable. Then, like something out of a bad dream, it began to snow. Big, wet flakes. As snow accumulated under darkening skies, the mood of discomfort in the schoolroom turned to fear. The teacher dismissed the students, directing them to find refuge immediately at the nearest house on their way. The barefoot students ran through the snow to a house they could see only a few hundred yards distant. But it was locked, and no one answered their hammering and cries. Drenched in the whirling snow, the four children felt the first gusts of panic. By this time, the schoolhouse was deserted, too. So, no going back.

With their survival now at stake, the nine-year-old boy, leader of the group, devised a complicated plan. Each of them would take a turn on his back with their feet in his jacket pockets while the others ran as far ahead as they could, stopping only when the cold of the snow became unbearable. He told them to rub each other's feet while they waited for him. And so they rotated, a hundred or so yards at a time, over the course of two miles of rocky, icy road, with snow rising up to their knees. At last in hailing distance of the first house, the girls were rescued by their startled father, leaping like a deer across the snowdrifts. The young hero of the story survived a while longer in the storm before he too found a place at the open fire in the cabin. Sitting too close, the pain of his thawing body overwhelmed his senses, and he fell unconscious. His feet had been torn to ribbons on the icy sticks and pebbles of his frozen march, and he could not walk for days.

So goes one of the innumerable tales of common suffering arising from the summer of 1816, the year Americans came to call "Eighteen-Hundred-and-Froze-to-Death." In a season marked by bizarre fluctuations of temperature, June 6—the day the Annsville schoolchildren found themselves caught in a snowstorm—stands as its surreal, wintry apex, an iconic day in the history of American weather. This unheard-of June snow, followed by other severe frosts through the summer, laid waste to staple crops and fruit stocks throughout the Atlantic states from Maine south even to the Carolinas. The cascading short- and medium-term impacts of this disaster—on food production, demography, and ultimately the entire U.S. economy—mark the 1816 summer as the most destructive extreme weather event of the nineteenth century.

The first signs of Tambora's doom-filled arrival on American shores came in early May. Newspapers in Washington, D.C., reported the sudden appearance of choking dust clouds over the capital: "the whole

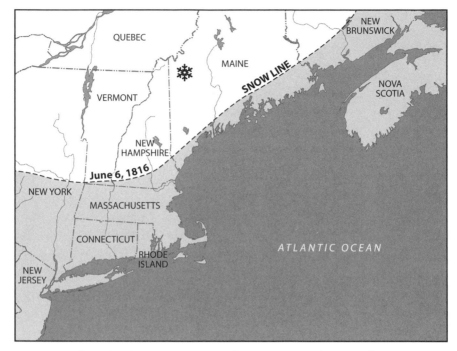

FIGURE 9.1. The volcanic summer storm of June 6, 1816—the most famous extreme weather event in nineteenth-century America—enveloped most of New England. The frigid system extended as far south as Bennington, Vermont, and Concord, New Hampshire, bringing snows even to the northern border of Massachusetts. (Adapted from Henry Stommel and Elizabeth Stommel, *Volcano Weather: The Story of 1816, the Year without a Summer* [Newport, RI: Seven Seas Press, 1983], 28–29.)

atmosphere is filled with a thick haze, the inconvenience of which is not diminished by the clouds of impalpable dust which float in the air."[4] The insinuating presence of Tambora's aerosol cloud dimmed the sun across the entire North Atlantic region, wreaking havoc on the evolution of seasonally benign weather systems. The rogue snowstorm that almost killed the schoolchildren in upstate New York originated with an intense, stationary high-pressure system off the east coast of Greenland in late May. This effectively blocked the eastward trajectory of North American weather, funneling Arctic air southward—a system characteristic of deep winter. As this cold air encountered the warmer atmosphere to the south through the week of May 28–June 4, it brought wildly unstable conditions to New England and Canada. An exaggerated temperature gradient across the mid-Atlantic latitudes intensified the overall energy of the emerging system.[5]

Extreme weather was now guaranteed. A depression passing across the Great Lakes stalled abruptly over Quebec, where it developed into a massive trough, sucking cold air into New England. A sudden cold snap was followed by two seasonable days in excess of 80°F, then widespread frosts on the night of June 3–4. In a normal year, such a night would mark an exceptional minimum temperature for June. But not in the Tambora year, 1816. Riding its unnaturally southerly jet stream, the cold front, mixing with warmer air above, brought icy precipitation to upstate New York and destructive thunderstorms to Pennsylvania. North of Harrisburg, a thousand acres of oats and rye were destroyed. Meanwhile, down in Virginia, pioneer meteorologist Thomas Jefferson recorded another dry day at his Monticello farm and worried about the effects of the developing drought on his fragile wheat crop.

On June 5, weather along the Atlantic seaboard turned full topsy-turvy. The sky turned black with hailstones at what is now Winston-Salem in North Carolina, while, to the rear of the storm track, Boston baked in temperatures over 90°. In the wake of this depression, however, the high-pressure system asserted itself once more, ushering in the frigid, northwest winds that would bring unwelcome Christmas snows to New England on June 6 and 7: "the most gloomy and extraordinary weather ever seen," in the words of one Vermonter.[6] Farmers who had planted their major crops in the spring had already experienced the dreaded "black frost" in mid-May. The cold wave of June 5–11 that dumped a foot or more of snow across the Northeast fully devastated their corn and grain fields. The region's orchards, where apple trees had only just blossomed, suffered massive losses. Birds fell stone dead out of trees, while farmers feared for the survival of their sheep, recently shorn. Frost conditions spread south to Richmond, Virginia, by June 9, and to the west as far as Cincinnati.

Had the June frosts been a singular event, resilient farmers along the East Coast might have prevailed with their second or third crops, and 1816 not passed into legend as the disastrous Year without a Summer. Through an anxious June, northern farmers prayed for deliverance.

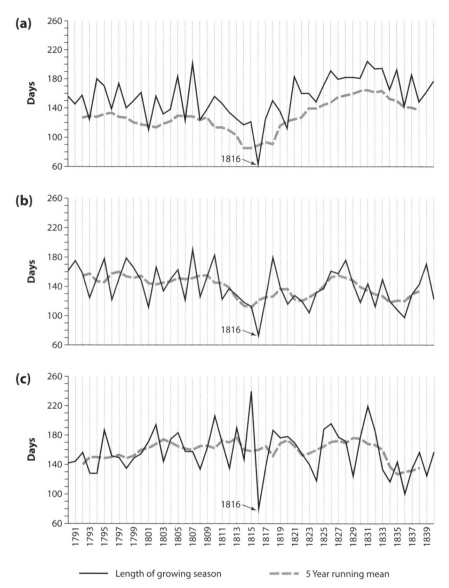

FIGURE 9.2. This graph, based on a fifty-year average, shows the impact of Tambora's Frankenstein weather on the New England growing season in 1816—for (a) southern Maine, (b) southern New Hampshire, and (c) eastern Massachusetts, respectively. The notorious conditions of the "Year without a Summer" cut the growing season by half or more. Abbreviated summer seasons across the volcanic decade of the 1810s are particularly evident here for Maine and New Hampshire. (C. R. Harington, ed., *The Year without a Summer? World Climate in 1816* [Ottawa: Canadian Museum of Nature, 1992], 133; Courtesy of the Canadian Museum of Nature.)

"Great frost," wrote Calvin Mansfield of Connecticut in his weather journal—"we must learn to be humble." But a frigid weather system similar to the early June snowstorm returned again in the first week of July. In the wake of this second disaster, the *New Hampshire Patriot* reported "fears of a general famine."[7] Killing frosts returned again in late August, blasting farmers' hopes a third time.

The year 1816 was not the coldest in the annals of the eastern United States in terms of average temperatures, but it is the only year to record frosts in each of the summer growing months, June, July, and August. Extreme dry conditions also prevailed. Newspapers reported the melancholy sight of "fields burned from drought." This deadly combination of frost and drought ensured the summer of 1816's notoriety as the shortest growing season in history—fewer than 70 days in New Haven, for example, as opposed to an average 126 days. Lack of hay meant feeding starving cattle with corn, reducing the overall supply. Adding to the misery, poor forest management since colonial times had resulted in dwindling reserves of first-growth timber throughout the Northeast and a shortage of firewood. Talk of an approaching famine intensified when a late September frost ended hopes of salvaging even a vestigial New England corn crop, while down south, planters in Georgia and South Carolina contemplated the loss of half their cotton harvest.[8]

For the retired third president of the United States, the disastrous summer of 1816 in the Atlantic states raised the specter of the drought of 1755, when scores of Virginians had died of outright starvation. In September, Jefferson reported to Albert Gallatin in Paris that "we have had the most extraordinary year of drought and cold ever known in the history of America."[9] Sparked by acute drought conditions, forest fires raged up and down the Atlantic states. As a winter wheat farmer, Jefferson felt the impact of the cold, dry summer most acutely the following year. In August 1817 he reported that for a second successive season "a great part of my own crop has not yielded seed," while his neighbors had taken to releasing their cattle onto the ruined wheat fields to make what they could of feeding on the stunted remains. Even three years later, the trend had not fully reversed. In 1820, Jefferson lamented to his steward that the successful planting of staple crops—wheat, tobacco, oats, corn—"seems [to] become more and more difficult every year."[10]

The cold summer of 1816, followed by the continuing poor harvest years of 1816–20, represented a personal crisis for Jefferson at many levels. In economic terms, he faced decisive ruin. His wheat crops withered in the frosts and drought, plunging him deeper into the longstanding debts from which he would never escape. At the moral and intellectual level, however, he faced perhaps an even greater crisis. If the weather in North America was actually getting *colder* as the years went by, and the climate less hospitable for agriculture, didn't the entire Jeffersonian argument for an agrarian republic come crumbling to the ground, not to mention westward expansion? In an 1817 letter, he expressed open concern about the fate of his Monticello farm "if the seasons should, against the course of nature hitherto observed, continue constantly hostile to our agriculture."[11] Selling up Monticello would be more than a personal embarrassment or economic hardship; it would represent the collapse of his romantic, lifelong vision for agricultural America. Must he now, as an old man, acknowledge that the European climate pessimists had been right in the end? That the United States—despite the heroic efforts of its yeoman-citizens in clearing and plowing the land—must be classified among the irredeemable places of the Earth, in the words of celebrated French scientist the Comte de Buffon, "des terres ingrates, froides, et denuées"—an ungrateful land, cold and barren?[12]

COLD NEW WORLD . . .

In the opening of his landmark essay "Des époques de la nature" (1778), Buffon describes the creation of the world in the collision of a comet with the sun. From this fiery beginning, our immolated planet gradually cooled, and was cooling still. In fact, the gradual refrigeration of the earth would necessarily continue—an idea that returned to haunt the Shelleys as they toured the Alps in 1816—until the Earth was "colder than ice" and bereft of life.[13] Until that long-distant date, however,

FIGURE 9.3. This portrait of Thomas Jefferson from 1821 captures the emotional traumas of the third president's final years, dominated by the destructive sequence of bad weather, crop failure, and economic turmoil that crippled the Atlantic states in the post-Tambora period. (© Thomas Jefferson Foundation at Monticello. Photo: Edward Owen.)

the originary heat emanating from the Earth's core provided the life-principle of the animate world by setting the temperature of its various regions. Climate, by Buffon's formulation, equaled temperature. And temperature, in turn, determined the relative "energy" and fecundity of nature around the globe.

FIGURE 9.4. A portrait of the Comte de Buffon early in his glittering career, to commemorate his 1753 election to the Académie Française. Fittingly—given Buffon's status as one of the last intellectual lions of the French ancien régime—the portrait hangs today in the royal palace of Versailles. (© RMN/Art Resource, New York.)

The great loser in this uneven thermal distribution, it turns out, was America. According to Buffon, "in such a situation is the continent of America placed, and so formed, that everything concurs to diminish the action of heat." Buffon's grand, multivolume *Histoire Naturelle* (1749–88) drew strong criticism from the French clergy for its scant references to a divine plan in nature. In the fledgling republic of the United States, however, controversy raged over Buffon's explicit anticolonial ecology, in particular the supposed "degeneration" of New World species under a relentlessly cold climate regime. From his study in the secluded village of Montbard—the headquarters of European natural science in the late eighteenth century—Buffon assessed the numbers, variety, and size of species collected from the distant Americas and concluded that in the New World "animated nature . . . is less active, less varied, and even less vigorous." This affliction extended to the native human population, whose menfolk lacked virile strength and failed to assert themselves over the wilderness: "Nature has withheld from them the most precious spark of her torch." Further proof of New World degeneration lay in the fact that animals transported from Europe to America failed to thrive there, becoming "shrivelled and diminished."[14]

Buffon belonged to the first generation of European naturalists with access to substantial samples of global flora and fauna. For such ambitious men of science—eager to subsume nature's bewildering variety within a single grand theory—climate supplied a most attractive explanatory power. For example, one needed only compare the mean temperatures of Quebec and Paris—located near the same line of latitude—to conclude that "in the new world there is much less heat and more moisture than in the old." Temperature alone, Buffon argued, had made the Americas a "perfect desert," where "the men are cold and the animals diminutive."[15] To promoters of the early republic, the most offensive implication of Buffon's argument was that Anglo-Saxon settlers of North America were destined to "degenerate" in the supposed manner of the native flora and fauna. And all because of a bit of cold weather! This serious controversy also offered abundant material for humorists. In the opening pages of his *Sketch Book* (1819), Washington Irving gave as his reason for touring Europe in 1815 an "earnest desire . . . to visit this land of wonders . . . and see the gigantic race from which I am degenerated."

Buffon's theories of New World degeneration appeared as ridiculous to Americans of the early republic as they do to us now. But their impact on the educated European public—as a theoretical bedrock of anticolonial opinion—was long-lasting. The image of North America as a cold and inhospitable continent, where nature could not thrive and Europeans ought not venture, became an *idée reçu*, to be rehearsed again and again in such texts as William Robertson's much-reprinted *History of America* (1788), where he translated Buffon's passages on American climate almost word for word. Buffon's influence even shows up in the radical poetry of the teenage Percy Shelley.[16] His first major poem, *Queen Mab* (1813), rehearses the Buffonian New World pessimism of the early 1760s, long settled as a mainstream European view:

> Man, where the gloom of the long polar night
> Lowers o'er the snow-clad rocks and frozen soil,
> Where scarce the hardiest herb that braves the frost
> Basks in the moonlight's ineffectual glow,
> Shrank with the plants, and darkened with the night;
> His chilled and narrow energies, his heart,
> Insensible to courage, truth, or love,
> His stunted stature and imbecile frame,
> Marked him for some abortion of the earth,
> Fit compeer of the bears that roamed around.[17]

—and so on (in unedifying strains). In this passage from *Queen Mab* it is easy to see how the Enlightenment anthropology of climate developed its reputation as the insidious intellectual precursor to nineteenth-century theories of race. If a self-proclaimed radical and humanitarian such as Shelley could believe this nonsense, who would not believe it? Even Mary Shelley was reading Buffon in 1817. In *Frankenstein*, her Creature, a wretched "abortion of the earth," naturally wishes to flee civilized Europe to the "degenerated" Americas.

The New World climate controversy is a mostly forgotten theme of early transatlantic relations.[18] But for Americans in London or Paris—who could not attend a dinner party without hearing the same old jokes about freezing weather and degeneration—Buffonian climate pessimism stood as a serious affront to patriotic pride. There could be no touchier subject for an American than the weather, except perhaps slavery. Opinion makers in Europe such as Buffon had tied the concept of climate so closely with culture that to complain of the American weather was to insult Americans themselves and to question the very viability of their infant republic.

Inevitably the founding fathers, especially those based for periods in Europe, found themselves enlisted as New World champions in the drawn-out battle over American climate. None took up the cudgels more aggressively than Thomas Jefferson, who, among his many distinctions, has been called the premier meteorologist of the early republic. In the early 1780s, Jefferson found himself the American ambassador to France. His official duties were light, so he occupied himself in reading all available literature on American geography and natural science, while circulating his celebrated *Notes on the State of Virginia* (first published in a French edition in 1785). A primary object of that book—a landmark in American geography—was to launch a full-fronted attack on Buffon's anti-American climate pessimism.

First, to demolish Buffon's argument for the physical diminution of New World animals, Jefferson pointed to the growing archaeological record: "It is well known that on the Ohio, and many parts of America further north, tusks, grinders, and skeletons of unparalleled magnitude, are found in great numbers." To lend empirical weight to Jefferson's arguments, *Notes* is studded with statistical tables comparing the size and weight of European animals with their New World equivalents, showing how American quadrupeds—including horses, bears, beavers, and flying squirrels—enjoyed a healthy superiority. As the coup de grâce, Jefferson offered his prime exhibit: the recently unearthed skeleton of a mammoth "six times the cubic volume of the elephant."[19] Because species extinction, to Jefferson's old-fashioned way of thinking, was impossible, living examples of this woolly giant would surely be found as Americans explored farther into the vast unknown territories of the West.

But it is in response to Buffon's argument that New World climate has presided over a degenerated *human* species that Jefferson becomes most passionate and oratorical. In a celebrated passage from *Notes on the State of Virginia*, the planter-politician rises to the defense of the embattled American Indian, whom Buffon had called "cold" and feeble. The influence of the icy climate, according to Buffon, extended even to the smallness of native genitals. Not so! Jefferson proclaims. The American native is physically powerful and "brave." His "friendships are strong and faithful" and his "ardor for the female" perfectly normal. Moreover, the climate has had no dulling effect on American intellect: the native "sensibility is keen . . . his vivacity and activity of mind is equal to ours."[20] The modern reader cannot help but be struck by the irony that a future president whose policy of Western expansion ensured the destruction of American Indian communities en masse should be so animated in their defense, albeit as "noble savages." But in championing the native American against Buffon, Jefferson was also defending his own Anglo-Saxon settler race against Buffon's remorseless climatic logic, which foretold "degeneration" for all migrants to the New World.

American climate optimism does not originate with Thomas Jefferson, of course. From Drayton's tribute to her "delicious land" in his "Ode to the Virginian Voyage" (1606) to St. John Crèvecoeur's popular *Letters from an American Farmer* (1782), Jefferson had a rich tradition of Arcadian imagery to call upon in evoking the hospitality of the American environment to human settlement. Virginia itself, in his own estimation, rated as the "Eden of the U.S.," while his wide reading in American travel literature taught him to think of the western United States—to the Mississippi and beyond—as a veritable cornucopia: "Here is health and joy, peace and plenty . . . the soil [is] excellent, the climate healthy and agreeable, and the winters moderate and short . . . no country in this quarter, if any in the world, is capable of larger or richer improvement than this."[21] Jefferson implicitly trusted such accounts because they underwrote his fondest hopes for a greater agrarian-based republic. The prosperity of the new nation and its unique personal freedoms depended, in Jefferson's imperial vision, on a near-infinite availability of sun-blessed fertile land to the West.

With Jefferson's attack on Buffon in *Notes on the State of Virginia*, the two contrary branches of colonial-era European writing on the Americas—one evoking Eden, the other a frozen tundra—converged in open conflict. When *Notes* appeared in print in 1785, Jefferson immediately sent a signed copy to the Comte de Buffon and arranged to visit his aged adversary at his provincial manor in Montbard to take up the debate in person. How fitting that these great armchair naturalists of the late eighteenth century should meet over dinner in a formal baroque dining room, eating off the finest imported chinaware, surrounded by leather-bound volumes of the Count's collected works. Undistracted by the real world, the two gentleman scientists were free to expound on their pet theories: that the Earth was made of glass (Buffon) or that gigantic mammoths and sloths roamed the American West under the loom of erupting volcanoes (Jefferson). But when it came to the matter of the undeclared war between them—on American climate and fauna—the septuagenarian Buffon proved an elusive opponent.

Jefferson had arrived at the Count's door carrying a large American panther skin, which Buffon once "had confounded with a cougar." Buffon instantly promised to correct his mistake in a new edition. This aristocratic concession with regard to detail merely laid the ground, however, for Buffon's denial of Jefferson on principle. He listened politely as his guest enumerated the pro-American arguments of *Notes on the State of Virginia* and made his passionate case for a benevolent New World climate. But Buffon did not deign to debate the young American ambassador. Instead, he merely reached for the latest hefty volume of his *Histoire Naturelle* and said, smilingly, "When Mr. Jefferson shall have read this, he will be perfectly satisfied that I am right."[22] For Buffon, the question of who had the larger genitals was never far from the surface.

No detailed account exists of the ensuing conversation, and Buffon died before their fascinating dialogue could be resumed. But we can well imagine their resorting to the usual tactic of ambitious men with sharply differing opinions who meet under conditions of politeness: they sought out subjects on which they could agree. Here, one strand of Buffonian climate theory overlapped appealingly with Jefferson's own: namely, that by transforming the natural landscape through agriculture, American settlers possessed the God-given power to radically improve their climate. In the utopian promise of climate engineering thus lay an "escape clause" for American patriots in the age of Buffon. One might respectfully disagree with one's continental friends on the natural advantages of the New World environment as it was "discovered" by Europeans because, once *settled*, the success of the American experiment lay in the hands of the colonists themselves. By deforesting and draining the land, by planting wheat and running cattle, American farmers were capable of transforming their "desert" land into "the most fruitful, healthy, and opulent in the world."[23]

Global cooling—slow but inexorable—remains the apocalyptic truth at the heart of Buffonian earth science.[24] Until the twentieth-century discovery of radioactivity, logic dictated as much. Emitting heat from its core since the hour of its molten creation, the Earth must eventually freeze. But Buffon, though a great controversialist, appears to have been uncomfortable with the mantle of prophet of doom. His theories should be more hopeful and offer something for everyone: hence his assurances that "Man is able to modify the influence of the climate that he inhabits—to fix, you might say, the temperature most convenient to him."[25]

The centrality of climate engineering to Buffon's most famous text explains how Jefferson could depart Montbard in 1785 expressing nothing but admiration for his host's "extraordinary powers" and "singularly agreeable" nature.[26] On the basis of the climate control argument that concludes "Des époques de la nature," Buffon was able to render a blueprint for agricultural imperialism utterly congenial to the Jeffersonian vision of America. The promise of climate salvation, like republican democracy itself, lay in agricultural expansion: "As the hallmark of his own civilization . . . Man changes the face of the earth, converts deserts into pastures, and wild heaths into fields of grain . . . and everywhere he produces abundance, there follows a great wave of migration; millions of men may inhabit the space formerly occupied by two or three hundred savages."[27] Europe had already enjoyed such a transformation. Now, the United States stood at the cusp of epochal environmental change and a similar population explosion. A bucolic western landscape, populated by waves of settler farmers, prosperous and free . . . just such a vision filled Jefferson's imagination two decades later when, in his dealings with another eminent Frenchman, he concluded the Louisiana Purchase.

Call it the Great Climate Compromise. One could complain about the weather in the new American republic, but all must agree it was changing for the better. The notion of an improving climate had been an article of faith in the American colonies since the days of Cotton Mather. A century later, climate boosterism still ruled the day. Constantin-François Volney, a brilliant geographer and friend of Jefferson's who traveled the United States in the early 1800s, grew accustomed to hearing the gospel of benevolent climate change preached wherever he went:

> I have collected similar testimonies in the whole course of my journies [*sic*]. . . . On the Ohio, at Gallipolis, Washington (Kentucky), Frankfort, Lexington, Cincinnati, Louisville, Niagara, Albany, everywhere the same changes have been mentioned and insisted on. Longer summers, later autumns, shorter winters, lighter and less lasting snows, and colds less violent were talked of by everybody; and these changes have always been described, in the newly settled districts not as gradual and slow, but as quick and sudden, in proportion to the extent of cultivation.[28]

Enlarging on this patriotic theme, the eminent New England naturalist Samuel Williams—who modeled his *Natural and Civil History of Vermont* (1794) on Jefferson's *Notes*—narrated in glowing detail the ameliorative powers of agriculture:

> When the settlers move into a new township their first business is to cut down the trees, clear up the lands, and sow them with grain. The earth is no sooner laid open to the influence of the sun and winds, than the effects of cultivation begin to appear. The surface of the earth becomes more warm and dry. As the settlements increase, these effects become more general and extensive: the cold decreases, the earth and air become more warm; and the whole temperature of the climate becomes more equal, uniform, and moderate . . . a remarkable change of this kind has been observed in all the settled parts of North America.[29]

Williams's arguments are not for marginal changes in weather but for *wholesale transformation* of the American climate regime brought about by European settlement. He estimates that agriculture has already warmed the inhabited parts of the country by 10°F (!) and that the benign, seaborne easterly winds, which formerly intruded no more than "thirty or forty miles," were now blowing merrily hundreds of miles inland to the slopes of the Appalachian Mountains.

In reading excited statements such as these—from the most respected scientific figures of the early republic—it seems impossible to overstate the giddy rush of climate optimism felt by educated Americans in the first decade of the nineteenth century. It was not to last. Ironically, a new, enlarged edition of Williams's *History of Vermont* appeared in 1809—the year the "Great Unknown" tropical eruption would usher in the coldest decade since European settlement. Temperature readings in New Haven, Connecticut, showed the years 1812–18 to be consistently 3–4°F below average.[30] Through the 1810s, the full-throated promotion of a warming climate by the American scientific elite met the chill reality of a national climate deterioration in which year after year posted a decline in average daytime temperatures and a sharp uptick in extreme weather events.

The impact of the volcanic decade of the 1810s was felt most keenly in New England. Williams's beloved Vermont, for example, suffered a dramatic population decline from which it did not recover for generations. This decade-long era of climate insecurity reached a chaotic climax in the Tamboran summer of 1816. Faced by that demoralizing natural disaster, the voices of the nationalist climate boosters fell conspicuously silent, while legions of ordinary folk sold up and went West—the first American climate change refugees. As one Connecticut resident remembered the crisis of 1816, "thousands feared or felt that New England was destined, henceforth, to become a part of the frigid zone."[31] The Buffonian nightmare was coming true.

TAMBORA AND THE PANIC OF 1819

For all the trauma of 1816's extreme weather, however, the twin pillars of Jefferson's worldview emerged from that meteorological disaster essentially intact. First, his climate optimism stood vindicated by the fact of the weather emergency's being temporary, however drastic. Normal temperature and precipitation patterns had resumed by the decade's end. Second, the salvific potential of the West shone brightly through the crisis. The frontier west of the Appalachians—"another Canaan" in Jefferson's eyes—was largely spared the monster cold fronts that repeatedly swept down from Canada across the Atlantic seaboard in the summer months of 1816.[32] Tree-ring studies that show "marked cooling" over western Europe and the eastern United States in 1816 and 1817 indicate no such arboreal stress in the West.[33] Such is the mix of cards dealt by climate change: the frontier, for the time being at least, profited wildly from the transatlantic subsistence crisis. Farming territories from the Ohio Valley to Illinois produced bumper crops and sold their bounty at record prices. The volume of flour and grain loaded on flatboats down the Mississippi to New Orleans in 1817 increased fourfold compared to 1814. The following summer, 1818, the boom continued unabated. A steamboat traveler heading north counted 643 flatboats bound for New Orleans in a brief few weeks' passage.[34] As the Atlantic states faced disastrous crop shortfalls, the new grain-producing regions of the west loomed more than ever as a Promised Land.

Thanks to the blessed intercession of western agriculture, the famine Jefferson himself had feared for Virginia and the Atlantic states in 1817–18 never eventuated, despite the general misery and alarm. What Jefferson did not reckon upon, however—and never properly understood—was the full ripple effect of "Eighteen-Hundred-and-Froze-to-Death" through the transatlantic economy and the primary role it played in America's first great depression of 1819–22.[35] As a cash-poor plantation farmer, Jefferson himself stood at the frontline of that impending economic collapse. Despite the impression of Olympian serenity he presented to visitors, Jefferson experienced the Tambora-driven economic crisis of those post-1816 years in a sickening spiral of indebtedness that culminated in the great personal humiliation of his old age: the mortgaging of his beloved Monticello.

By the late summer of 1818, the directors of the national bank of the United States had grown increasingly fearful of a credit bubble spurred by the extraordinary wave of migration and land speculation in the frontier West. A combination of 1816's disastrous weather and the subsequent demand for western grain on both sides of the Atlantic had lured settlers across the Appalachians in the tens of thousands. "So terrible was the year 1816," one New Hampshire diarist recorded, "that the people grew disheartened and many sold out and went west."[36]

Land speculators advertised aggressively for new settlers, convincing many New England farmers to sell up on the spot. Entire communities traveled together in wagon trains, looking to establish new townships in Ohio or Indiana; the latter welcomed forty-two thousand new inhabitants in 1816 alone.[37] "Old America seems to be breaking up and moving westward," wrote the English promoter Morris Birkbeck in his best-selling Illinois travelogue of 1818. One witness in Missouri rated the hordes of migrants on a scale of natural disasters, as a "flood," a "mountain torrent," an "avalanche." Jefferson himself called the migratory wave out of Virginia in 1816–18 "beyond anything imaginable."[38]

The desperate poor were sometimes reduced to crossing the mountains on foot, including a family of eight who trudged from Maine to Easton, Pennsylvania, arriving in the dead of winter with their smallest child nearly frozen in a hand cart.[39] But the call of Westward Ho! was heeded not only by impoverished refugees. According to influential travel writer Henry Fearon, who toured the length and breadth of the United States in 1817–18, the migrant wave also included "men of capital, of industry . . . who apprehended approaching evils" and were concerned "to provide for the future support and prosperity of their offspring."[40]

Western politicians took advantage of the economic distress in the East following the poor harvests of 1816 and 1817 by painting the advantages of trans-Appalachian migration in the gaudiest Jeffersonian colors: "There neither is, nor, in the nature of things can there ever be, anything like poverty there. All is ease, tranquility, and comfort," wrote one Missouri representative. Economic refugees from Europe's disastrous 1816 joined the tide. Despite the efforts of the British Parliament to limit the number of passengers aboard U.S.-bound vessels, 1817 witnessed the greatest number of British and European migrants yet to arrive on American soil in a single year. Official records were not kept in the days of the early republic, but even during the French Terror of the mid-1790s, annual migration from Europe to the United States never reached 10,000. In 1817, it spiked to 22,240.[41] To this must be added the British migrants who teemed south across the border from British Canada: 10,000 in 1817, swelling to 14,500 in 1818.[42] Few, it seems, intended to settle in the established cities of the Northeast. "I can scarcely walk a square," a Philadelphian reported, "without meeting with Irish, Dutch, English, and Scotch emigrants, whose destination is principally Ohio and Indiana." All of which meant bad news for the American Indians, who continued to be forced west to make way for the new arrivals. As a consequence of the first major westward expansion in U.S. history, a flurry of "treaties" formalized the ejection of indigenous peoples from lands they had inhabited time out of mind.

At the height of America's first real estate bubble, land prices out west replaced the weather as the "perpetual topic of conversation."[43] Makeshift banks sprung up overnight to handle the deluge of new settlers and their land claims. "The children of Israel could scarcely have presented a more motley array," recalled pioneering geographer Henry Schoolcraft, who traveled with the legions of migrants along the rivers to the interior. Expectations were running high: "To judge by the tone of general conversation, they meant, in their generation, to plough the Mississippi Valley from its head to its foot. . . . What a world of golden dreams was there!"[44] People called it "Ohio fever," which one no sooner recovered from than "Missouri fever" or "Illinois fever" took hold. Before the fever finally broke with the Panic of 1819, five new states had been established: Illinois, Indiana, Kentucky, Alabama, and, controversially, Missouri in 1820.

The land rush of 1817–18 lacked one crucial ingredient, however: capital. No national currency was in place to fund the large-scale movement of population across state lines into "empty" territory. Moreover, the hard currency Americans depended on—mostly Spanish silver—was simultaneously flowing in the opposite direction toward Europe to pay for the glut of imported British manufactures following the end of the war and to retire national debt. People fled west as money fled east.

With specie scarce and no established financial institutions across the Appalachians, the frontier land drive ran on confidence alone, through the circulation of notes of credit backed by personal endorsements.

In his letters, Thomas Jefferson said little of his own financial woes but predicted disaster for the country. The new national bank, to which he had essentially mortgaged his property, had been created on April 10, 1816—the ill-omened first anniversary of Tambora's eruption. Within a few short years, its haphazard policies would help trigger an implosion across the entire U.S. economy. To finance the land boom, the light-headed directors of the bank issued notes in excess of $22 million against a specie reserve of barely $2 million—at a time when the average farm laborer earned about $10 a month. America in 1817 was, as one pamphleteer later put it, a country of "paper gold, and paper land, and paper houses, and paper revenues, and paper government."[45]

This paper edifice of prosperity duly went up in flames. In July 1818 the national bank, wary of collapse, directed its branches to sharply reduce the issue of notes. Adding to the sense of crisis, the bank announced that its own loans would be renewed in the future only at a reduction of 12½ percent, with the debtor liable for the balance. Jefferson was with James Madison, about to meet with fellow trustees of the planned University of Virginia, when he got the news. It came upon him, he wrote that day, like "a clap of thunder."[46] He had no cash to cover the depreciated notes. There was nothing to do but seek out more friends and family, and have them endorse more loans. Tormented with anxiety about the mounting financial crises he faced, Jefferson developed a painful outbreak of boils in August. His hopes for relief rested with selling his flour at high prices at the Richmond market.

But now, in September 1818, the other shoe dropped. In an ironic twist to the Tambora tale, it was the return of good weather that doomed the U.S. economy in the post-volcanic period. In 1817, the total value of U.S. grain exports had doubled in value, to $23 million, while the following year Britain imported the greatest volume of grain in its history. Over half a million barrels of American flour passed through the Liverpool docks.[47] But with the aerosol dust from Tambora's eruption washed from the North Atlantic atmosphere by the summer of 1818, normal crop-growing conditions returned to the European continent, which enjoyed a return of bumper harvests.

The poet Keats, whose brother George had recently joined the tide of British emigrants to the American West, celebrated the return of sunny English harvesting in his famous ode "To Autumn":

> Season of mists and mellow fruitfulness,
> Close bosom-friend of the maturing sun;
> Conspiring with him how to load and bless
> With fruit the vines that round the thatch-eves run (ll. 1–4)

The sumptuous happiness of that poem speaks for the relief of an entire continent. But English "mellow fruitfulness" turned sour on their American brothers across the pond, including Keats's own, who returned penniless to London during the Panic (only to raid his tubercular sibling's trust fund and bolt back to Kentucky). As a result of improved European crop yields, the market for American commodities plummeted overnight. Philadelphia wheat went into free fall from a high of $3.11 a bushel in 1817, bottoming out at 82 cents in 1821. A hundredweight of Virginia flour, worth over $14 at the height of the boom, brought only $4 in 1820.[48] Meanwhile, cotton prices fell by two-thirds, devastating the economy of the Tennessee Valley. When Jefferson's flour finally reached the Richmond market, it returned half the price he expected. By September, he was describing to friends the total collapse of his health. It's not difficult to see the agitating cause. "To owe what one cannot pay," he wrote at the time, "is a constant torment."[49]

By early 1819, the collapse of European demand for American cotton and grain, combined with the national bank's contraction of credit, had generated a full-blown financial panic in the United States. In February, with the price of flour fallen still further, Jefferson made what must have been a heartbreaking decision: he would sell portions of his Monticello property to pay his debts. Yet greater humiliation followed. In April, he was forced to admit to his agent Gibson that his lands had little market value. The value of property had fallen to less than a year's rent—and with no buyers to be had even at that price! This latest

Location	1815	1816	1817	1818	1819	1820
United States	100	124	154	127	86	59
Britain	100	117	146	131	114	102
France	100	145	185	126	94	98
Switzerland	100	162	235	121	81	75
Austria (Vienna)	100	188	183	52	30	40
Bavaria (Munich)	100	190	301	131	n.a.	46
Unweighted averages	100	154	201	115	81	70

FIGURE 9.5. Index of wholesale grain prices for western Europe and the United States, 1815–20, which shows the dramatic surge in transatlantic grain prices owing to the ruined harvests of 1816 and 1817, followed by their equally stunning collapse in 1819–20. Index numbers for the United States, Britain, France, and Switzerland represent wheat prices, and Austria (Vienna) and Bavaria (Munich) rye prices. Adapted from John D. Post, *The Last Great Subsistence Crisis in the Western World* (Baltimore: Johns Hopkins University Press, 1977), 37.

setback saw Jefferson take on a further debt load totaling almost $11,000, consisting of five separate loans from three banks. His "entanglements," as Jefferson called them, were now beyond any rational scale. The former president, along with millions of his fellow citizens, was floating on the rim of a giant bubble—reality fast receding beneath him: "All is confusion, uncertainty, and panic."[50]

Thomas Jefferson experienced the Panic of 1819 as a profound, personal depression of body and mind. "After two years of prostrate health," he wrote in the fall of 1820, "you have the old, infirm, and nerveless body I now am, unable to write but with pain, and unwilling to think without necessity."[51] In one sense, Jefferson's pain in these years is purely personal and pitiable: the unfortunate result of an old man's bad luck and wavering judgment. At another level, however, it stands as a poignant symbol of national distress. Jefferson was never closer to the citizenry he ruled for eight years—and in whose imagination he held such a vaunted place—than during the Panic of 1819, when over three hundred banks across the nation failed overnight. His experience of crop failure, debt, and humiliation after 1816 was shared by legions of

FIGURE 9.6. This idyllic view of Monticello was drawn the summer before the incumbent's death in 1826. It shows Jefferson's grandchildren gamboling over the estate's famous gardens. An unspoken irony of the image (in addition to the absence of slaves) is that Jefferson's debts—blown out beyond all recovery by the Panic of 1819—ensured that his descendants would not inherit Monticello. The children are enjoying summer at their grandfather's estate for the last time. (© Thomas Jefferson Foundation at Monticello; Photo: H. Andrew Johnson.)

his fellow Americans who lost their farms, jobs, homes, and life savings in an economic crisis of epic dimensions. "Never were such hard times," Jefferson wrote in April 1820 as the U.S. economy subsided from panic into a general depression. "Not a dollar is passing from one to another." The former president was well aware of the scope of the suffering, both nationally and near at hand. With the people of Virginia "in a condition of unparalleled distress," a breakdown of civil order seemed imminent. "I fear," he wrote, "local insurrections against these horrible sacrifices of property."[52]

In the words of historian Daniel Dupre, "The Panic of 1819 is not just a tale of banks and currency, of debtors and creditors and sheriff's sales; it is the story of an emotional upheaval and shattering of a collective expectation of progress and prosperity."[53] The raw statistics are brutal enough. In the crash year 1819, imports dropped 55% and exports more than 40% as business nationwide ground to a halt. In the West, newspapers likened the subsequent depression to the "Famine of Egypt." In Kentucky, debtors overflowed the jails and filled the town squares. Court records from the period are conspicuously missing from frontier towns of that state, raising the possibility they were later destroyed to hide community shame at the wave of criminal default brought on by the Panic. Some bankrupt men were driven to suicide, while others hid in churches or escaped into the wilderness with their families, hoping to restart their lives beyond the reach of debt collectors.

A ghost map of early nineteenth-century Missouri would show dozens of towns—with hopeful names such as Missouriton, Washington, and Monticello—founded during the heyday of the boom, only to be abandoned in the crash.[54] "Last year we talked of the difficulties of paying for our lands," wrote one speculator, "this year the question is, how to exist." Bankrupt land investors in the West and South ultimately forfeited hundreds of thousands of acres back to the federal government, stalling westward migration for more than a decade. Meanwhile, in the eastern cities, John Quincy Adams reported "enormous numbers of persons utterly ruined; multitudes in deep distress; and a general mass disaffection to the government."[55] Outside Baltimore, the indigent set up shantytowns, as would become a common sight in the American cities of the 1930s. In Boston, the streets "present[ed] a dull and uncheery spectacle—silence reigns, [with] gloom and despondency in every countenance."[56] The Panic of 1819 had succeeded to the unrelenting misery of "hard times."

The eruption of Tambora was thus responsible for far more than a single summerless year in New England. By precipitating violent short-term fluctuations in the price of grain in the transatlantic region, the Tamboran weather of 1815–18 was a primary cause of the United States' first major economic depression.[57] The collapse of the European market for American commodities in 1818, combined with a currency crisis, crippled every sector of the U.S. economy and brought an abrupt end to the so-called Era of Good Feelings that followed the end of the war with Britain. The financial meltdown of 1819–22—when the U.S.

population stood at barely ten million—might not have equaled the
Great Depression of the 1930s in scale, but it was the young republic's
first depression and, by many measures, the most wrenching national
economic crisis of the nineteenth century. The Tambora period and its
aftermath introduced shell-shocked Americans to the nasty vicissitudes
of both climate and commerce. It dealt a body blow to republican optimism, the effects of which reverberated well into the 1820s.

Furthermore, from an environmental viewpoint, the legacy of the
mid- to late 1810s—endless cold weather followed by proverbial "hard
times"—challenged Americans' Jeffersonian faith that the country's
future lay in the sunny conjunction of an improving climate, expanding agriculture, and economic growth. The coincidence of hard times
with the return of good weather after 1818 was especially demoralizing.
A frustrated Committee on Manufactures in Washington wondered at
the paradox of financial collapse in a country where "the sea, the forest, the earth, yield their abundance; the labor of man is rewarded;
pestilence, famine, or war commit no ravages; no calamity has visited
the people; peace smiles on us: [and] plenty blesses the land." In the
Panic of 1819, the American public experienced the apparent historical
decoupling of climate and prosperity, a signature of industrial modernity. While the Year without a Summer made beggars of the Europeans,
it was the *return* of normal weather in 1818 that threw the dependent
American economy into disarray.[58] The United States had suffered the
first rude shock of nineteenth-century economic globalization.

Post-panic, the United States moved quickly to improve its competi-
tive position in the global market. Just as the extreme weather disaster of
1816 set off a wave of transport infrastructure construction—including
the Cumberland Pike and Erie Canal—to better connect the country's
agriculture with its points of sale, response to the 1819–22 depression
focused on the rationalization of currency and depersonalization of
credit. No more signing extravagant, unbacked bills for friends, family,
and creditors, à la Thomas Jefferson. Instead, a new Hamiltonian era of
national credit and currency evolved under a more stably administered
national bank—until Andrew Jackson arrived to tear it all down again
in the 1830s.[59]

In all this, however, one truth about the Tambora emergency in the transatlantic zone should not be lost sight of. Without an international market in grain to ship food supplies from Baltimore, Constantinople, and Odessa to the capitals of western Europe, millions more people would surely have starved to death in the European subsistence crisis of 1816–17.[60] The crippling depression that subsequently afflicted the United States represents the economic "downside" of this market response to a short-term international food emergency. By 1818, Tambora's pall of gloom had passed, as abruptly and mysteriously as it had come. And yet its destructive impact—both ecological and economic—endured many years. During the subsequent "hard times" of 1819–22, American farmers abandoned their golden wheat fields in the West, while warehouses full of grain—so recently in desperate demand—were left to rot unsold in the port of New Orleans.

THE RETURN OF CLIMATE PESSIMISM

For all their differences, Thomas Jefferson and the Comte de Buffon shared the Enlightenment dream of geo-engineering—of a global climate moderated and perfected through agricultural science. Little did they realize, however, that a massive global experiment in climate engineering was just underway in their lifetimes through the industrial application of fossil fuels. Fast forward two centuries since Tambora's eruption, we find a world in which cheap energy has brought enormous wealth and comfort to the populations of many countries and pulled untold millions out of poverty. But carbon-driven modernity has come at great ecological cost. Carbon waste—unregulated and unpriced—continues to alter the essential chemistry of the atmosphere and oceans. As a result, farmers of the twenty-first century face regional changes in weather patterns at a scale and pace human beings have not seen since the first emergence of agriculture ten thousand years ago.

The weather of our third millennium, as altered by humankind, is in fact heading in the opposite direction to that foreseen by Jefferson and Buffon. Instead of a managed, ameliorative warming of the colder regions of the Earth, we now have a runaway climate system increasingly prone to unpredictable extremes of drought, flood, and storms—something like the Frankenstein's weather of 1816–17. Summers are increasingly hot rather than cold, but the net negative impact on agriculture is the same: declining yields and escalating prices on world commodity markets. For more than a century before the birth of Jefferson, American farmers had worked to adapt their agriculture to the prevailing weather regimes across the United States. Now they face the daunting task of maintaining their crop yields in a deteriorating, unstable climate, exactly the predicament the nation's agricultural system faced in the Tamboran decade of the 1810s.

Debates over American climate—epitomized by the Jefferson-Buffon exchange—took center stage in the political life of the early republic. Moreover, the 1816 weather disaster furthered the sciences of climate and meteorology in the United States, at least in terms of its status as a professional discipline and office of government. It is no coincidence that the earliest meteorological journal kept by the Army Medical Department dates from July 1816, in the midst of a national weather emergency. In 1817, as a response to general concern over climate deterioration, the federal Land Office likewise ordered its twenty regional branches to begin systematic records of temperature and precipitation. The following year, the army followed suit under the supervision of the surgeon general. A new, federalized era of meteorological data gathering had begun.[61]

Ironically, however, as American meteorology became more professionalized, weather awareness receded among the general population. The earlier certainties of an ameliorating climate had vanished post-1816, to be replaced by lukewarm doubt and indifference. By 1825, the author of the national *Meteorological Register* reported a mere muddle of opinion on this once galvanizing issue: "some [contend] that as the population increased and civilization extended the climate became warmer, others that it became colder, and others that there was no change."[62] The daily preoccupation with weather conditions and seasonality—a hallmark of the Jeffersonian age—ebbed as an ever-decreasing fraction of the population lived and worked on the land and the United States developed into the metropolitan-based manufacturing society Jefferson foresaw with such clear-eyed loathing. After 1820, climate change likewise faded from the transatlantic conversation. The United States did not need a moderate climate—or the promise of one—to attract investment and build industries in the new era. Undergirding this industrial development, a modern financial system evolved that, for the most part, insulated ordinary Americans from the vicissitudes of weather and crop yields.[63]

One consequence of the long decline of American rural life has been a profound climate illiteracy among the political class, which reflects that of the citizenry at large. So the American public was ill-prepared when, in the mid-1980s—two hundred years after Jefferson's famous dinner with the Comte de Buffon—climate change returned as a live issue in American politics. As in the age of Jefferson, climate change continues to be a subject peculiarly fired with controversy, passion, and no small amount of ignorance. Twenty-first-century Americans, like their predecessors in the early republic, take climate very personally. As this chapter has shown, climate pessimism has never been popular in the United States, whose citizens long ago embraced Jefferson's sunny optimism at the expense of the icy prognoses of the Comte de Buffon. To suggest that the American climate is bad or getting worse is, in this historical sense, unpatriotic.

But patriotic or not, Buffonian climate pessimism now enjoys its modern revival, in drag, as global warming. The twenty-first-century climate emergency, as we all know, involves not lack of heat but too much, while higher volumes of water vapor destabilize the carbon-charged atmosphere. A new era of Frankenstein's weather—heat waves, droughts, wild storms, and floods—are increasingly part of the fabric of American life. Cascading extreme weather events, of ever-greater size and frequency, now loom as a serious threat to the agriculture and prosperity of the nation—its Jeffersonian core. No longer an historical footnote, climate pessimism has returned in full, bounding and ferocious, like the dog who will have its day at last.

ET IN EXTREMIS EGO

So near to us in geological time, 1815–18 was—in human terms—a remote age of small farms and horses. Railroads and steamships lay a few decades in the future, the mass-produced automobile a century away. Cocooned within our own advanced food and transport infrastructure in the developed world, it is easy to lose sight of the significance of animal mortality on a preindustrial world. The cold conditions and lack of grain in Europe during the 1816–18 climate crisis meant untold losses in livestock and the death of countless thousands of horses. The decimation of Europe's preindustrial transport system prompted one young German nobleman, Karl von Drais, to tinker with alternative modes of locomotion. The result: a "velocipede," a crude prototype of the modern bicycle.[1]

This historical vignette from the Tambora period tells us three things. First, that this book has contained almost nothing on the perhaps irrecoverable history of Tambora's impact on the world's animal and biotic populations. Second, that the downstream effects of an ecological crisis on the scale of Tambora evolve over many decades (a commercial, mass-produced bicycle wasn't developed until the 1890s). And third, that these long-term impacts of drastic climate deterioration are felt not only in the physical world—in human and animal suffering—but in the world of ideas and technology, where sudden environmental dislocations act as an extraordinary stimulant.

The eruption of Mount Tambora in 1815 was a natural disaster with a long dragon-like tail. As a geological act of global climate sabotage, its explosion was responsible, in ways direct and indirect, for a Shakespearean shelf's worth of human tragedies: from transcontinental famines, to a global cholera epidemic, to an exponential growth in the Chinese opium trade, to the first "great" depression in U.S. history. But there are parallel stories of human progress also arising from the Tambora crisis. Among those examined in this book have been the first speculative steps toward Ice Age theory; a golden age of Arctic literature, science, and navigation (despite the body count); a great leap forward in the sciences of agriculture and meteorology (the first weather map!); and the tentative emergence of a modern, liberal idea of the state, in which government's responsibilities expanded, in the minds of many, to include the welfare of its citizens in times of crisis. In terms of progressive idea generation, we can only wish that our current climate emergency will do the same, that the pressures of painful adaptation to increasing extreme weather conditions will produce grand solutions to climate change itself. And soon.

Many such solutions have already been proposed, one of which bears directly on the subject of this book, namely geo-engineered moderation of the world's warming climate. Since Nobel Prize–winning earth scientist Paul Crutzen brought new publicity to the idea in a 2006 article, a flurry of scientific papers has considered the possibility of imitating the cooling impact of volcanic eruptions on global temperature by artificially injecting massive volumes of sulfate aerosols into the stratosphere.[2]

Evidence for the extraordinary folly of this idea is to be found on every page of this book. How ironic if a series of Tambora-sized eruptions were—by some dubious miracle of international cooperation—to be authorized, and we were to relive the weather chaos of 1815–18 not as a natural disaster but a staged volcanic catastrophe (à la Surrey Gardens) of our own making? On the other hand, for a technology-rich civilization addicted to photoshop realities and spectacular special effects—and little versed in the teleconnections of climate and human society—a supersized artificial volcano blast might make for an appropriate climax . . . or finale. But in such cynical reflections, as King Lear warns us, "madness lies."

Further complicating the project of artificial aerosol cooling is recent evidence that naturally occurring volcanic eruptions are already having an intermittent chilling effect on the impacts of man-made global warming. Even four relatively minor eruptions in the 2000–2010 period—in Indonesia, Ecuador, Papua New Guinea, and Montserrat—increased the stratospheric aerosol load sufficient to offset man-made warming by 25%.[3] Unfortunately, however, there is no volcanic solution—natural or artificial—to twenty-first-century climate change, since volcanoes explode unpredictably, while human carbon emissions remain on an inexorable upward trajectory.

• • •

A celebrated seventeenth-century painting by the landscapist Nicolas Poussin shows shepherds of antiquity in an idyllic rural setting under temperate blue skies, clustered about a stone tomb. They are pondering the enigmatic inscription on the tomb, which reads "Et in Arcadia Ego." Death appears to be saying: "I am with you here, even in paradise." The virile shepherds and a beautiful, well-dressed woman look on the mausoleum with curiosity and respect—and a healthy dose of disbelief. But life is so good here! they seem to say. The painting portrays the inevitability of death even as it celebrates, in loving rendering of the Arcadian landscape, the equable, life-sustaining climate that vitalized human civilization in the Mediterranean for thousands of years. Scientists call it the Holocene—the roughly twelve millennia of temperature stability since the last glacial epoch during which human agriculture flourished and with it human arts and culture. The cultural wealth of our Holocene legacy is embodied in Poussin's exquisite painting itself, even as its serene atmosphere offers us a window onto a now vanished climatological past.

It was also Paul Crutzen, would-be climate engineer, who popularized the term "Anthropocene" to distinguish the benevolent Holocene of Poussin's Arcadia from our modern fossil-fuel era. The

FIGURE E.1. Nicolas Poussin's *Shepherds of Arcadia* (1637–38), better known as *Et in Arcadia Ego* after the inscription barely legible on the tomb. Poussin's exact message in his painting—is he emphasizing the reality of death or the pleasures of Arcadia?—has been a debate point among art historians for centuries. In our century of climate change, the "choice" between these interpretations captures perfectly the situation facing global humanity. (© Musée du Louvre/Art Resource, New York.)

Anthropocene designates the age of accelerated urban-industrial development and population increase since 1750 that has wrought ever-increasing changes on the Earth's biosphere, and witnessed the contamination and depletion of its natural resources.[4] Where the deep time modulations of geology or Darwinian natural selection might be said to have long ruled unchallenged, we are now at the point, Crutzen and others have argued, where human beings have assumed the principal levers of control over biophysical conditions on planet Earth.

The Anthropocene came often to my mind as I sailed through open Arctic waters in the summer of 2012, on my intermittent global quest to retrace, in sublunary fashion, the course of Tambora's volcanic plume from the tropics to the poles. On a journey of some 2,500 nautical miles from the west coast of Greenland to northwestern Canada, our sturdy

Russian arctic vessel encountered a solitary thin film of sea ice.[5] The cruise ships and oil companies had already moved in, while the few polar bears we saw prowled along sandy beaches. I had come to see The Arctic!—but was already too late. This was not the smooth-running air-conditioner of our Holocene planet as I should have found it but a new Arctic altogether—the doomed-to-be-ice-free Arctic of the Anthropocene. When the summer ice does disappear altogether—a decade or two from now—the Arctic will be an open sea for the first time in a million years. Lounging on deck in a light cotton jacket, it was difficult to imagine the terrible privations of Ross or Franklin's crews in these waters: beset for years, frostbitten, starving. But for the flowering Arctic tundra on the shore, we might have been cruising the Hudson or the Rhine. "Et in Arcadia ego," I thought.

Because if death is always with us, according to Poussin—even in sunny, serene Arcadia— how much more true is this of a weather world in extremis, such as the Tambora-driven hell of 1815–18? The global death toll from Tambora was likely in the millions—or tens of millions if we include the worldwide cholera epidemic its eruption almost certainly triggered. Reflecting on Tambora's three-year extreme weather regime of the early nineteenth century, we cannot predict what the mortality rates of our own extended climate emergency will be in the Anthropocene age. Advanced infrastructure, technology, and communications guarantee first-world citizens far greater resilience against weather disasters than that available to the global peasantry of 1815–18. On the other hand, the unsustainable consumption rates of the West and high population growth in developing countries increase humanity's vulnerability overall. Already, the stresses on human food, water, and public health systems worldwide from a changing climate are immense and dangerously unpredictable.

Whatever the historical variables separating 1815 and 2015, the Tambora case study allows us, at the very least, to appreciate the scale of the current threat to human civilization. If a three-year climate change event in the early 1800s was capable of such destruction and of shaping human affairs to the extent I have described in this book, then the future impacts of multidecadal climate change must be truly off the charts. It's difficult to think of any aspect of our lives and societies that won't be transformed in the coming decades, and for the worse.

Tambora's history, then, like Mary Shelley's *Frankenstein*, ultimately tells a cautionary tale. Both warn against the technological hubris of our modernity through figures of intense and widespread suffering. "All men hate the wretched," laments the Creature, Shelley's literary projection of the homeless, starving poor of Europe in Tambora's aftermath.[6] Two centuries on, the global ranks of the wretched are set to increase exponentially in coming decades at the hands of our own climate "Frankenstein," a monster who feeds on carbon waste and grows more violent by the year. Failure to draw down the carbon emissions and rampant deforestation that drive climate change brings us ever closer to the traumatized world of 1815–18 writ large. Imagine, if you can bear it, the "Seven Sorrows" of Tambora scaled to a planet of ten billion, and lasting for centuries. While the cautionary tale of Tambora's eruption has here been told at last, to write a history of the global climate breakdown of our early third millennium—with its relentless attrition of human well-being across all latitudes—might be beyond the capacity of any future historian, whose chaotic world it is.

ACKNOWLEDGMENTS

My grateful thanks, first of all, to my editor at Princeton, Ingrid Gner-
lich, for her enthusiastic support for this project. She never flinched
from the challenge of our highwire crossing between the sciences and
humanities and has been a consistent and generous advocate. If our
interdisciplinary ambitions have been realized here, others at Princeton
University Press likewise share in that success, including Alison Kalett,
Eric Henney, Debbie Tegarden, and Jennifer Backer—consummate pro-
fessionals all.

At the outset of my Tambora research, indispensable help in grap-
pling with the scientific literature on Tambora came from climatologist and Intergovernmental Panel on Climate Change (IPCC) coauthor
Don Wuebbles and his graduate student Darienne Curio-Sanchez.
Other scientists in the fields of volcanology and climatology provided
illumination at crucial points, including Michael Schlesinger, Stephen
Self, Bob Rauber, and anonymous readers at Princeton University Press.
The guidance of all the above was vital for a humanist entering into the
sometimes difficult terrains (and atmospheres) of the physical sciences.

The mostly unglamorous detective work required to bring the
global *Tambora* story to light involved tracking down a vast range of
nineteenth-century sources across multiple continents. This could not
have been done without the magnificent library at the University of
Illinois at my disposal and a host of expert staff. I wish to thank, in particular, Shuyong Jiang of the Asian Library, Adam Doskey in the Rare
Book Room, and the staff of the Interlibrary Loan Office who dealt

with my intermittent hailstorm of requests with great patience and efficiency. For research tips on "The Panic of 1819," richly layered with moral support, my sincere thanks to David Brady in Springfield, Illinois.

As a recovery mission in early nineteenth-century global history, the writing of this book involved significant transnational challenges, of which China was the greatest for me as a nonspecialist. I take the opportunity here to thank the librarians at Yunnan University for their hospitality and aid during my research trip in 2011, Professor Rong Guangqi of Wuhan University for his first draft translations of Li Yuyang's (now) unforgettable poetry, and a cohort of graduate students in Asian studies at Illinois who assisted at various stages of the Yunnan famine research project. For the Bengali side of the story, gratitude goes to my colleague Anustup Basu for his introduction to cholera folklore. And last, but by no means least, for the transformative experience of climbing Mount Tambora itself, I wish to thank my guide, Ma-cho, the crew from Lombok, and my Sumbawan hosts at various stations on the journey.

Final thanks and love go to my wife, Nancy, and the children I left at home on my various Tamboran wanderings (Sumbawa, India, China, the Arctic . . .). The quality of souvenirs I brought home for Lucas and Clara was spotty at best, but this book of many fathers will, I hope, compensate in time for the absence and other deficiencies of the one.

NOTES

INTRODUCTION
FRANKENSTEIN'S WEATHER

1. Benjamin Franklin, "Meteorological Imaginations and Conjectures," *Memoirs of the Literary and Philosophical Society of Manchester* 2 (1784): 357–61; see also *Works of Benjamin Franklin*, ed. John Bigelow (New York, 1888), 8:486–89.

2. *The Philosophical Magazine and Journal* 46 (July–December 1815): 231.

3. C. S. Zerefos et al., "Atmospheric Effects of Volcanic Eruptions as Seen by Famous Artists and Depicted in Their Paintings," *Atmospheric Chemistry and Physics* 7 (2007): 4027–42.

4. Janet Todd, *Death and the Maidens: Fanny Wollstonecraft and the Shelley Circle* (Berkeley: Counterpoint, 2007), xiii.

5. The standard popular transatlantic history—Henry Stommel and Elizabeth Stommel's *Volcano Weather: The Story of 1816, the Year without a Summer* (Newport, RI: Seven Seas Press, 1983)—has recently been superseded by a more detailed narrative, *The Year without Summer: 1816 and the Volcano That Darkened the World and Changed History* (New York: St. Martin's, 2013), by William K. Klingaman and Nicholas P. Klingaman. The latter has the advantage of greater meteorological detail, though its geographical range and social themes do not extend beyond the terms set by John D. Post's marvelous scholarly history, *The Last Great Subsistence Crisis in the Western World* (Baltimore: Johns Hopkins University Press, 1977).

CHAPTER ONE
THE POMPEII OF THE EAST

1. Quoted in Adam Zamoyski, *Rites of Peace: The Fall of Napoleon and the Congress of Vienna* (New York: Harper Collins, 2007), 477.

2. Bernice de Jong Boers, "The 'Arab' of the Indonesian Archipelago: The Famed Horse Breeds of Sumbawa," in *Breeds of Empire: The "Invention" of the Horse in Southern Africa and Maritime South East Asia, 1500–1950*, ed. Greg Bankoff and Sandra Swart (Copenhagen: NIAS Press, 2009), 51–64.

3. Bernice de Jong Boers, "A Volcanic Eruption in Indonesia and Its Aftermath," *Indonesia* 60 (October 1995): 37–60.

4. Charles Assey, *On the Trade to China, and the Indian Archipelago, with Observations on the Insecurity of British Interests in That Quarter* (London, 1819), 13–14. For a recent historical account of indigenous piracy in the region, see James F. Warren, *Iranun and Balangingi: Globalization, Maritime Raiding and the Birth of Ethnicity* (Singapore: Singapore University Press, 2002), and note 5 below.

5. James F. Warren, "A Tale of Two Centuries: The Globalization of Maritime Raiding and Piracy in South East Asia at the End of the Eighteenth and Twentieth Centuries," in *A World of Water: Rain, Rivers, and Seas in South-East Asian Histories*, ed. Peter Boomgaard (Leiden: KITLV Press, 2001), 133.

6. John Crawfurd, *A Descriptive Dictionary of the Indian Islands and Adjacent Countries* (New York: Oxford University Press, 1971), 437.

7. M. J. Hitchcock, "Is This Evidence for the Lost Kingdoms of Tambora?" *Indonesia Circle* 33 (1984): 34.

8. Bernice de Jong Boers, 38.

9. Jeyamalar Kathirithamby-Wells, "Socio-political Structures and the South-East Asian Ecosystem: An Historical Perspective up to the Mid-Nineteenth Century," in *Asian Perceptions of Nature: A Critical Approach*, ed. Ole Bruun and Arne Kalland (Richmond: Curzon Press, 1995), 27.

10. The most accessible account of the eruption that consolidates scientific and historical research on Tambora is found in Clive Oppenheimer, *Eruptions That Shook the World* (New York: Cambridge University Press, 2011), 295–319. The volcanological description of the eruption that follows, here and in chapter 2, is likewise drawn from the following key sources in the scientific literature: Stephen Self et al., "Volcanological Study of the Great Tambora Eruption of 1815," *Geology* 12 (November 1984): 659–63; Richard B. Stothers, "The Great Tambora Eruption in 1815 and Its Aftermath," *Science* 224 (June 15, 1984): 1191–98; J. Foden, "The Petrology of Tambora Volcano: A Model for the 1815 Eruption," *Journal of Volcanology and Geothermal Research* 27 (1986): 1–41; Haraldur Sigurdsson and Steven Carey, "Plinian and Co-ignimbrite Tephra Fall from the 1815 Eruption of Tambora Volcano," *Bulletin of Volcanology* 51 (1989): 243–70; Sigurdsson and Carey, "Eruptive History of Tambora Volcano," *Mitteilungen Geologisch-Palaeontologische Institut* (University of Hamburg) 70 (1992): 187–206; Stephen Self, Ralf Gertisser, et al., "Magma Volume, Volatile Emissions, and Stratospheric Aerosols from the 1815 Eruption of Tambora," *Geophysical Research Letters* 31 (2004): L20608; and Ralf Gertisser, Stephen Self, et al., "Processes and Timescales of Magma Genesis and Differentiation Leading to the Great Tambora Eruption in 1815," *Journal of Petrology* 53.2 (2012): 271–97.

11. J. T. Ross, "Narrative of the Effects of the Eruption from the Tomboro [*sic*] Mountain in the Island of Sumbawa," *Batavian Transactions* 8 (1816): 1–25. This collection of eyewitness reports, commissioned by Raffles, constitutes the sole contemporary account of the eruption.

12. The epic poem from Bima from which these lines are taken was first transcribed in Malay about 1830. The full poem appears in a modern French edition: Henri-Chambert Loir, ed., *Syair Kerajaan Bima* (Jakarta: Ecole Française D'Extreme-Orient, 1982). The stanza on Tambora quoted here appears in Boers, "A Volcanic Eruption," 37 [translation modified].

13. "'Pompeii of the East' Discovered," BBC News, February 28, 2006. The volcanologist Haraldur Sigurdsson coined the phrase "Pompeii of the East," prompting a flurry of media reports.

14. Estimates have varied greatly. Oppenheimer sets the likely range at 60,000–120,000 victims of the eruption, including those who succumbed to disease and starvation in its immediate aftermath (*Eruptions That Shook the World*, 311).

15. In the words of one contemporary Javanese chronicler: "Awe-inspiring (*langkung huébat*) to behold, they are as though protected by the very angels (*lir pinayungan malékat*) and they strike terror into men's hearts." Peter Carey, ed., *The British in Java, 1811–1816: A Javanese Account* (New York: Oxford University Press, 1992), 17, 79.

16. Ross, "Narrative of the Effects of the Eruption," 3–4, 13.

17. Ibid., 14–15.

18. J. H. Moor, *Notices of the Indian Archipelago and Adjacent Countries* (Singapore, 1837), 95.

19. Ross, "Narrative of the Effects of the Eruption," 9.

20. Peter R. Goethals, *Aspects of Local Government in a Sumbawan Village* (Ithaca: Cornell University Department of Far Eastern Studies, 1961), 19.

21. Raffles, at the encouragement of his superior in Calcutta, Lord Minto, interpreted the Slave Felony Act of 1811 as prohibiting the import of slaves to any British possession, a regulation he tightened in the period leading up to the eruption. See H.R.C. Wright, "Raffles and the Slave Trade at Batavia in 1812," *Historical Journal* 3.2 (1960): 184–91; Susan Abeyasekere, "Slaves in Batavia: Insights from a Slave Register," in *Slavery, Bondage, and Dependency in Southeast Asia*, ed. Anthony Reid (New York: St. Martin's, 1993), 289; and Gillen D'Arcy Wood, "The Volcano Lover: Climate, Colonialism and Slavery in Raffles's *History of Java*," *Journal of Early Modern Cultural Studies* 8.2 (2008): 33–54.

22. Boers, "A Volcanic Eruption," 49.

23. Quoted in ibid., 47.

24. Roseanne D'Arrigo et al., "Monsoon Drought over Java, Indonesia, during the Past Two Centuries," *Geophysical Research Letters* 33 (2006): L04709.

25. Boers, "A Volcanic Eruption," 38.

26. *Jakarta Globe*, October 17, 2012.

27. Hitchcock, "Is This Evidence for the Lost Kingdoms of Tambora?" 30–33.

28. My narrative of this legend is drawn from conversations and interviews conducted with Sumbawans and residents of surrounding islands during my visit there in March 2011.

29. Munshi Abdullah, *Autobiography*, trans. W. G. Shellabear (Singapore, 1918), 51–52.

30. Sir Thomas Stamford Raffles, *The History of Java*, intro. John Bastin (New York: Oxford University Press, 1965), 1:7.

31. Ibid., 1:119.

32. Adam Smith, *The Wealth of Nations* (New York: Modern Library, 1937), 602.

33. In the pungent phraseology of his Dedication to the Prince of Wales, Raffles's goals for the British administration in Java were "to uphold the weak, to put down lawless force, to lighten the chain of slave . . . to promote the arts, sciences, and literature, to establish humane institutions." See C. E. Wurtzburg's comprehensive account of Raffles's career, *Raffles of the Eastern Isles*, 2nd ed. (New York: Oxford University Press, 1984).

CHAPTER TWO
THE LITTLE (VOLCANIC) ICE AGE

1. "Ode to Naples" (1820), l. 1.

2. *The Letters of Percy Bysshe Shelley*, ed. Frederick L. Jones (Oxford: Clarendon Press, 1964), 2:491.

3. Alexander von Humboldt, *Personal Narrative of Travels to the Equinoctial Regions of the New Continent during the Years 1799–1804*, trans. Helen Maria Williams (London, 1815); George Steuart Mackenzie, *Travels in the Island of Iceland, during the Summer of 1810* (London, 1811), 111–12.

4. Humphry Davy, "On the Phenomena of Volcanoes," in *Collected Works*, ed. John Davy (London, 1840), 6:346.

5. Madame de Staël, *Corinne; or, Italy*, trans. Emily Baldwin and Paulina Driver (London, 1906), 224.

6. Arch Johnston and N. K. Moran, "John Weisman Accounts of the Earthquakes" (Memphis: Center for Earthquake Research and Information), http://www.ceri.memphis.edu/compendium/eyewitness/wiseman.html.

7. Richard Altick, *The Shows of London* (Cambridge, MA: Harvard University Press, 1978), 96.

8. See Nicholas Daly, "The Volcanic Disaster Narrative: From Pleasure Garden to Canvas, Page, and Stage." *Victorian Studies* 53.2 (Winter 2011): 255–85.

9. See Mary Ashburn Miller, "Mountain, Become a Volcano: The Image of the Volcano in the Rhetoric of the French Revolution," *French Historical Studies* 32.4 (Fall 2009): 555–85.

10. See Richard B. Alley, *The Two-Mile Time Machine: Ice Cores, Abrupt Climate Change, and Our Future* (Princeton: Princeton University Press, 2000).

11. Jihong Dai et al., "Ice Core Evidence for an Explosive Tropical Volcanic Eruption 6 Years Preceding Tambora," *Journal of Geophysical Research* 96 (September 1991): 17361–66.

12. Ellen Mosley-Thompson et al., "High Resolution Ice Core Records of Late Holocene Volcanism: Current and Future Contributions from the Greenland PARCA Cores," *Volcanism and the Earth's Atmosphere* 139 (2003): 153–64. A 2006 paper, based on ice core results from the Yukon that showed chemically distinct 1809 tephra deposits from those found at the poles, offers an alternative scenario, namely two separate 1809 eruptions: a major tropical eruption coinciding with a minor northern hemisphere eruption. If substantiated, this would necessarily have implications for the modeling of 1809's climatic impact. See Kaplan Yalcin et al., "Ice Core Evidence for a Second Volcanic Eruption around 1809 in the Northern Hemisphere," *Geophysical Research Letters* 33 (2006): L14706.

13. Luke Howard, *The Climate of London: Deduced from Meteorological Observations, Made at Different Places, in the Neighbourhood of the Metropolis* (London, 1820), vol. 1, table 44, n.p.

14. A. V. Eliseev and I. I. Mokhov, "Influence of Volcanic Activity on Climate Change in the Past Several Centuries: Assessments with a Climate Model of Intermediate Complexity," *Izvestiya: Atmospheric and Oceanic Physics* 44.6 (2008): 671–83.

15. Jihong Cole-Dai et al., "Cold Decade (AD 1810–19) Caused by Tambora (1815) and Another (1809) Stratospheric Volcanic Eruption," *Geophysical Research Letters* 36 (2009): L22703; Rosanne D'Arrigo et al., "The Impact of Volcanic Forcing on Tropical Temperatures during the Past Four Centuries," *Nature Geoscience* 2 (2009): 51–56.

16. George Mackenzie, *The System of the Weather of the British Islands* (Edinburgh, 1821), table VII, n.p.

17. Samuel Taylor Coleridge, "Fancy in Nubibus; or, the Poet in the Clouds," *Blackwood's Magazine* (November 1819), l. 9.

18. See Gillen D'Arcy Wood, "Clouds, Constable, Climate Change," *Wordsworth Circle* 38.1–2 (2007): 25–34.

19. See *Nature Geoscience*, October 9, 2011 [press release].

20. Alan Robock, "The 'Little Ice Age': Northern Hemisphere Average Observations and Model Calculations," *Science* 206 (December 1979): 1402–4. For a more thorough rebuttal of the solar variability theory, see G. Hegerl et al., "Detection of Volcanic, Solar, and Greenhouse Signals in Paleo-Reconstructions of Northern Hemisphere Temperature," *Geophysical Research Letters* 30.5 (2003): 1242. On the volcanic source of the "Dalton Minimum" period, see Sebastian Wagner and Eduardo Zorita, "The Influence of Volcanic, Solar, and CO_2 Forcing on the Temperatures in the Dalton Minimum (1790–1830): A Model Study," *Climate Dynamics* 225 (2005): 205–18.

21. Thomas J. Crowley et al., "Volcanism and the Little Ice Age," *PAGES News* 16.2 (April 2008): 22–23.

22. As the first eruption observed with modern scientific instruments, Pinatubo has been vital to research on connections between tropical volcanism and global climate, and hence on the development of scholarship on Tambora.

23. Christopher G. Newhall and Stephen Self, "The Volcanic Explosivity Index (VEI): An Estimate of Explosive Magnitude for Historical Volcanism," *Journal of Geophysical Research* 87 (February 1982): 1231–38; Lee Siebert, Tom Simkin, and Paul Kimberley, *Volcanoes of the World*, 3rd ed. (Washington, DC: Smithsonian Institution, 2010).

24. See Alan Robock and Melissa P. Free, "Ice Cores as an Index of Global Volcanism from 1850 to the Present," *Journal of Geophysical Research* 100 (June 1995): 11549–67, and the recent follow-up paper coauthored by Robock: Chaochao Gao et al., "Volcanic Forcing of Climate over the Past 1500 Years: An Improved Ice Core–Based Index for Climate Models," *Journal of Geophysical Research* 113 (2008): D23111.

25. For a summary and development of the theory of Santorini's biblical connections, see Barbara J. Sivertsen, *The Parting of the Sea: How Volcanoes, Earthquakes, and Plagues Shaped the Story of the Exodus* (Princeton: Princeton University Press, 2009).

CHAPTER THREE
"THIS END OF THE WORLD WEATHER"

1. *The Letters of Mary Wollstonecraft Shelley*, ed. Betty B. Bennett (Baltimore: Johns Hopkins University Press, 1980), 1:17; *The Clairmont Correspondence: Letters of Claire Clairmont, Charles Clairmont, and Fanny Imlay Godwin*, ed. Marion Kingston Stocking (Baltimore: Johns Hopkins University Press, 1995), 1:48. That same wet summer in nearby Hampshire, Jane Austen, housebound and ailing, was writing her late masterpiece, *Persuasion*, in which the melancholy heroine must reconcile herself to a Tambora-style future with no "second spring."

2. *Letters of Mary Shelley*, 1:20.

3. Paul Henchoz, "L'Année de la Misère (1816–17) dans la Région de Montreux," *Revue Historique Vaudoise* 42 (1934): 72.

4. Post, *The Last Great Subsistence Crisis in the Western World*, 21.

5. Henchoz, "L'Année de la Misère," 83.

6. *Childe Harold's Pilgrimage*, III:860–61, 864–66, 873–77.

7. H. H. Lamb and A. I. Johnson, *Secular Variations of the Atmospheric Circulation since 1750* (London: Her Majesty's Stationery Office, 1966), 117–21. See also Drew Shindell et al., "Dynamic Winter Climate Response to Large Tropical Volcanic Eruptions since 1600," *Journal of Geophysical Research* 109 (2004): D05104.

8. David P. Schneider et al., "Climate Response to Large, High-Latitude and Low-Latitude Volcanic Eruptions in the Community Climate System Model," *Journal of Geophysical Research* 114 (2009): D15101; E. M. Fischer et al., "European Climate Response to Tropical Volcanic Eruptions over the Last Half Millennium," *Geophysical Research Letters* 34 (2007): L05707.

9. Alastair G. Dawson et al., "A 200-Year Record of Gale Frequency, Edinburgh, Scotland: Possible Link with High-Magnitude Volcanic Eruptions," *The Holocene* 7.3 (1997): 337–41.

10. *The Diary of Dr. John William Polidori, 1816: Relating to Byron, Shelley, etc.*, ed. William Michael Rossetti (London, 1911), 127–8.

11. *The Prose Works of Percy Bysshe Shelley*, ed. E. B. Murray (Oxford: Clarendon Press, 1993), 1:284.

12. For an excellent account of the impact of the Genevan weather on the Shelley Circle in 1816, and on the writing of *Frankenstein*, see John Clubbe, "The Tempest-toss'd Summer of 1816: Mary Shelley's *Frankenstein*," in *Literature in Context*, ed. Joachim Schwend et al. (Frankfurt: Peter Lang, 1992), 219–36. For another reading of weather in *Frankenstein* (that makes allusion to the Tambora eruption), see Bill Phillips, "*Frankenstein* and Mary Shelley's 'Wet, Ungenial Summer,'" *Atlantis* 28.2 (December 2006): 59–68.

13. *Letters of Percy Bysshe Shelley*, ed. Jones, 1:483–84; *Byron's Letters and Journals*, ed. Leslie A. Marchand (Cambridge, MA: Harvard University Press, 1976), 5:82.

14. Shelley, *Frankenstein*, 40.

15. Jeff Masters, "2010–2011: Earth's Most Extreme Weather since 1816?" *Weather Underground*, June 24, 2011, www.wunderground.com.

16. Thomas Forster, *Researches into Atmospheric Phaenomena* (London, 1813), vii [my emphasis].

17. For an accessible modern account of Howard's career, the essay on clouds, and its influence, see Richard Hamblyn, *The Invention of Clouds: How an Amateur Meteorologist Forged the Language of the Skies* (New York: Farrar, Straus, and Giroux, 2001).

18. Howard, *Climate of London*, 2:112.

19. Quotes in this section are drawn from ibid., vol. 1, tables 120–26, n.p.

20. Ibid., 2:109.

21. Karl Schneider-Carius, *Weather Science, Weather Research: A History of Their Problems and Findings from Documents during Three Thousand Years*, trans. from German (New Delhi, 1975), 179.

22. Heinrich Brandes, *Beiträgen zur Witterungskunde* (Leipzig, 1820).

23. Howard, *Climate of London*, 2:46–47.

24. Ibid., 2:55, 58, 62, 73.

25. Carl von Clausewitz, *Politische Schriften und Briefe*, ed. Hans Rothfels (Munich: Drei Masken Verlag, 1992), 190 [my translation].

26. *The Times*, January 1, 1817.

27. See A. J. Peacock, *Bread or Blood: A Study of the Agrarian Riots in East Anglia in 1816* (London: Victor Gollancz, 1965).

28. *The Times*, March 4, 1817.

29. Post, *The Last Great Subsistence Crisis*, 94.

30. Marc Henrioud, "L'Année de la Misère en Suisse et plus particulièrement dans le Canton de Vaud," *Revue Historique Vaudoise* 25 (1917): 137–38, 120.

31. Post, *The Last Great Subsistence Crisis*, 39.

32. Thomas Raffles, *Letters during a Tour through Some Parts of France, Savoy, Switzerland, Germany, and the Netherlands, in the Summer of 1817* (London, 1818), 145, 156.

33. Henrioud, "L'Année de la Misère en Suisse," 139, 117.

34. Post, *The Last Great Subsistence Crisis*, 51, 128.

35. Quoted in ibid., 80.

36. Henchoz, "L'Année de la Misère," 81.

37. William J. Bromwell, *History of Immigration to the United States [1819–55] . . . with an Introductory Review of the Progress and Extent of Immigration to the United States prior to 1819* (New York, 1856), 15.

38. Henrioud, "L'Année de la Misère en Suisse," 142; Post, *The Last Great Subsistence Crisis*, 165.

39. *Letters of Percy Bysshe Shelley*, 1:480–81.

40. Mary Shelley, *Frankenstein*, ed. Maurice Hindle (New York: Penguin, 1992), 9.

41. Henry Matthews, *Diary of an Invalid* (London, 1824), 192–93.

42. Ibid., 103.

43. *Byron's Letters and Journals*, ed. Marchand, 5:86.

44. *His Very Self and Voice: Collected Conversations of Lord Byron*, ed. Ernest J. Lovell (New York: Macmillan, 1954), 299.

45. *London Chronicle*, July 23, 1816.

46. Post, *The Last Great Subsistence Crisis*, 25.

47. See Jeffrey Vail's rich account in " 'The bright sun was extinguish'd': The Bologna Prophecy and Byron's 'Darkness,'" *Wordsworth Circle* 28.3 (Summer 1997): 183–92.

48. *London Chronicle*, July 23, 1816; *Morning Chronicle*, June 21, 1816.

49. *Quarterly Journal of Science and the Arts* 2 (1817): 420.

50. *London Chronicle*, July 23, 1816; Henrioud, "L'Année de la Misère en Suisse," 119.

51. Henchoz, "L'Année de la Misère," 74.

52. *Morning Chronicle*, July 18, 1816; *The Examiner*, July 21, 1816.

53. Henchoz, "L'Année de la Misère," 75.

54. Samuel Taylor Coleridge, *Collected Letters*, ed. Earl Leslie Griggs (Oxford: Clarendon Press, 2000), 4:660.

CHAPTER FOUR
BLUE DEATH IN BENGAL

1. Homer, *The Iliad*, trans. Richmond Lattimore (Chicago: University of Chicago Press, 1951), 1:44–52.

2. Frederick Corbyn, *A Treatise on the Epidemic Cholera, as it has Prevailed in India* (Calcutta, 1832); see also James Jameson, *Report on the Epidemick Cholera Morbus, As It Visited the Territories Subject to the Presidency of Bengal in the Years 1817, 1818, and 1819* (Calcutta, 1820), 12–19.

3. David Arnold, *Colonizing the Body: State Medicine and Epidemic Disease in Nineteenth-Century India* (Berkeley: University of California Press, 1993), 171.

4. Jameson, *Report on the Epidemick Cholera Morbus*, 17; Marquess of Hastings, *Private Journal*, ed. Marchioness of Bute (London, 1858), 2:241.

5. Jameson, *Report on the Epidemick Cholera Morbus*, 18.

6. Thomas Medwin, *The Angler in Wales; or, Days and Nights of Sportsmen* (London, 1834), 2:346.

7. See Ernest J. Lovell's biography of Medwin, *Captain Medwin: Friend of Byron and Shelley* (Austin: University of Texas Press, 1962). As the subtitle suggests, Medwin's own literary fame rests on his role as chronicler of the lives and opinion of his more gifted contemporaries, in particular his authorship of the instantly notorious, best-selling *Conversations of Lord Byron* (1824).

8. Lovell, *Captain Medwin*, 74.

9. J. J. Higginbotham, ed., *Selections from the Asiatic Journal and Monthly Register* (Madras, 1875), 14.

10. Ulrich von Rad et al., "A 5000-yr Record of Climate Change in Varved Sediments from the Oxygen Minimum Zone off Pakistan, Northeastern Arabian Sea," *Quaternary Research* 51.1 (1999): 43.

11. Alan Robock, "Volcanic Eruptions and Climate," *Reviews of Geophysics* 38.2 (May 2000): 191–219.

12. See Bin Wang, ed., *The Asian Monsoon* (Berlin: Springer, 2006); and P. K. Das, *The Monsoons: A Perspective* (New Delhi: Indian National Science Academy, 1984).

13. Descriptions of the weather in Bengal in the Tambora period are drawn from Jameson, *Report on the Epidemick Cholera Morbus*.

14. Schneider et al., "Climate Response to Large, High-Latitude and Low-Latitude Volcanic Eruptions in the Community Climate System Model," D15101.

15. Jameson, *Report on the Epidemick Cholera Morbus*, xliii.

16. Ram R. Yadav, "Basin Specificity of Climate Change in Western Himalaya, India: Tree-Ring Evidences," *Current Science* 92.10 (May 2007): 1424–29.

17. Jameson, *Report on the Epidemick Cholera Morbus*, xlviii.

18. For a description of the religious practices surrounding Ola Bibi, see Sunder Lal Hora, "Worship of the Deities, *Olā, Jholū*, and *Bŏn Bībī* in Lower Bengal," *Journal and Proceedings of the Asiatic Society of Bengal* 29 (1933): 1–4.

19. Jameson, *Report on the Epidemick Cholera Morbus*, lvi.

20. Emma Roberts, *Scenes and Characteristics of Hindostan* (London, 1835), 1:270.

21. James Statham, *Indian Recollections* (London, 1832), 214.

22. Quotations from documents housed in the India Office Records collection, British Library: "Proceedings Relative to the Measures which were Successfully Adopted with the View of Counteracting the Fatal Epidemic Denominated Cholera Morbus which Raged in the Town and Suburbs of Calcutta" [R. Leny, Secretary to the Medical Board, September 29, 1817]; "Papers of George Green Spilsbury," September 25, 1819; "Proceedings Relative . . ." [Robert Tytler, Assistant Surgeon at Jessore, October 1, 1817]; "Judicial Letters: Measures Adopted for Affording Medical Aid to the Persons Attacked with the Epidemic Disorder at Bewar, Banda and Cawpore" [Secretary to the Medical Board to W. B. Bayley, September 22, 1817]; James Jameson to Military Department, October 9, 1818; Board of Revenue to Fort George, November 28, 1818.

23. A. Moreau de Jonnés, *Rapport au Conseil Supérieur de Santé sur le cholera-morbus pestilentiel* (Paris, 1831).

24. For recent histories and cultural analysis of the 1817 cholera in India, see Arnold, *Colonizing the Body*, and Mark Harrison, *Climates and Constitutions: Health, Race, Environment and British Imperialism in India, 1600–1850* (New Delhi: Oxford University Press, 1999). For a global perspective on cholera, on which I have significantly relied in this chapter, see Christopher Hamlin, *Cholera: The Biography* (New York: Oxford University Press, 2009).

25. A selective bibliography of recent cholera research, on which my discussion is based, includes the following key publications: Luigi Vezzulli et al., "Environmental Reservoirs of *Vibrio cholerae* and Their Role in Cholera," *Environmental Microbiology Reports* 2.1 (2010): 27–33; Timothy E. Ford et al., "Using Satellite Images of Environmental Changes to Predict Infectious Disease Outbreaks," *Emerging Infectious Diseases* 15.9 (September 2009): 1341–46; Jane N. Zuckerman et al., "The True Burden and Risk of Cholera: Implications for Prevention and Control," *Lancet: Infectious Diseases* 7 (2007): 521–30; Mercedes Pascual et al., "Hyperinfectivity in Cholera: A New Mechanism for an Old Epidemiological Model?" *Public Library of Science* 3.6 (June 2006): 931–32; Katia Koelle et al., "Refractory Periods and Climate Forcing in Cholera Dynamics," *Nature* 436 (August 4, 2005): 696–700; Mercedes Pascual and Katia Koelle, "Disentangling Extrinsic from Intrinsic Factors in Disease Dynamics: A Nonlinear Time Series Approach with an Application to Cholera," *American Naturalist* 163.6 (June 2004): 901–13; Rita R. Colwell, "Infectious Disease and Environment: Cholera as a Paradigm for Waterborne Disease," *Perspectives: International Microbiology* 7 (2004): 285–89; G. Uma et al., "Recent Advances in Cholera Genetics," *Current Science* 85.11 (December 10, 2003): 1538–45; Kathryn L. Cottingham et al., "Environmental Microbe and Human Pathogen: The Ecology and Microbiology of *Vibrio Cholerae*," *Frontiers in Ecology and the Environment* 1.2 (2003): 80–86; Andrew E. Collins, "Vulnerability to Coastal Cholera Ecology," *Social Science and Medicine* 57 (2003): 1397–1407; Mercedes Pascual et al., "Cholera and Climate: Revisiting the Quantitative Evidence," *Microbes and Infection* 4 (2002): 237–45; Joachim Reidl and Karl E. Klose, "*Vibrio cholerae* and Cholera: Out of the Water and into the Host," *FEMS Microbiology Reviews* 26 (2002): 125–39; Xavier Rodó et al., "ENSO and Cholera: A Nonstationary Link Related to Climate Change?" *Proceedings of the National Academy of Sciences* 99.20 (October 1, 2002): 12901–6; Rita R. Colwell and Anwar Huq, "Marine Ecosystems and Cholera," *Hydrobiologia* 460 (2001): 141–45, 1341–46; Paul Shears, "Recent Developments in Cholera," *Current Opinion in Infectious Diseases* 14 (2001): 553–58; Mercedes Pascual and Menno Jan Bouma, "Seasonal and Interannual Cycles of Endemic Cholera in Bengal 1891–1940 in Relation to Climate and Geography," *Hydrobiologia* 460 (2001): 147–56; Mercedes Pascual et al., "Cholera Dynamics and El Niño-Southern Oscillation," *Science* 289 (September 8, 2000): 1766–69; Brad Lobitz et al., "Climate and Infectious Disease: Use of Remote Sensing for Detection of *Vibrio cholerae* by Indirect Measurement," *Proceedings of the National Academy of Science* 97.4 (February 15, 2000): 1438–43; and Rita R. Colwell, "Global Climate and Infectious Disease: The Cholera Paradigm," *Science* 274.5295 (December 20, 1996): 2025–31.

26. J. F. Heidelberg, "DNA Sequence of Both Chromosomes of the Cholera Pathogen Vibrio Cholerae," *Nature* 406.6795 (August 3, 2000): 477–83.

27. Pascual et al., "Cholera and Climate," 237.

28. Mark Harrison, *Disease and the Modern World: 1500 to the Present Day* (Cambridge: Polity Press, 2004), 105.

29. Hamlin, *Cholera: The Biography*, 97.

30. Ibid., 4.

CHAPTER FIVE
THE SEVEN SORROWS OF YUNNAN

1. Clarke Abel, *Narrative of a Journey in the Interior of China . . . in the Years 1816 and 1817* (London, 1818), 11–12.

2. Peiyuan Zhang, "Extraction of Climate Information from Chinese Historical Writings," *Late Imperial China* 14.2 (December 1993): 96–106.

3. The inspiration for this chapter lies with the original research of Yang Yuda of the University of Fudan in Shanghai, in particular his lengthy article coauthored with Man Zhmin and Zheng Jinyun, "The Great Yunnan Famine of the Jiaqing Period (1815–17) and the Eruption of the Volcano Mount Tambora," *Fudan Journal: Social Sciences* (2005): 79–85 (unpublished translation by Nicholas Brown). Two other recent Chinese sources provided supplementary information: Shuji Cao, "The Tambora Eruption and Chinese Social History," *Academics in China* 5 (September 2009): 37–41 (unpublished translation by Fang Wan); and Bantian Chen, "The Serious Famine in Yunnan during the Jiaqing Period (1796–1820)," *Shi Jie Bo Lan* [*Global Vision*] 8 (2010): 76–79 (unpublished translation by Fang Wan).

4. Bin Yang, *Between Wind and Clouds: The Making of Yunnan* (New York: Columbia University Press, 2009).

5. Bao Chenglan, ed., *Synoptic Meteorology in China* (Beijing: China Ocean Press, 1987); Jiacheng Zhang and Zhiguang Lin, *Climate of China*, trans. Ding Tan (New York: Wiley & Sons, 1992).

6. C. Patterson Giersch, *The Transformation of Qing China's Yunnan Frontier* (Cambridge, MA: Harvard University Press, 2006); see also Mark Elvin and Liu Ts'uijung, *Sediments of Time: Environment and Society in Chinese History* (New York: Cambridge University Press, 1998).

7. Giersch, *The Transformation of Qing China's Yunnan Frontier*, 145.

8. *Weather and Rice* (Laguna, Philippines: International Rice Research Institute, 1987).

9. Luigi Mariani et al., "Space and Time Behavior of Climatic Hazard of Low Temperature for Single Rice Crop in the Mid Latitude," *International Journal of Climatology* 29 (2009): 1863; *Science of the Rice Plant* (Tokyo: Food and Agriculture Policy Research Center, 1993–96).

10. Thomas H. C. Lee, *Education in Traditional China* (Boston: Brill, 2000).

11. A brief official biography of Li Yuyang is found in Ma Yao, *Yunnan Jian Shi* [A Short History of Yunnan] (Kunming: Yunnan ren min chu ban she, 1983), 404–6. A far richer portrait is available in his highly personal poetry.

12. Translation of select poems of Li Yuyang represents a collaborative effort over several years. My co-translator, Professor Rong Guangqi of Wuhan University, undertook the first literal translation of the texts. Shuyong Jiang, of the East Asian Library at the University of Illinois, identified the relevant poems while Fang Wan, Lingling Yao, Jing Chen, and Nicholas Brown assisted and/or translated contextual documents. The poems are translated from the original Chinese texts collected and reprinted as "Ji yuan shi chao," in *Cong Shu Ji Cheng*, vol. 178 (Xinwenfeng chu ban gong si, 1989), 1–103.

13. Robert Marks, *Tigers, Rice, Silk, and Silt: Environment and Economy in Late Imperial South China* (New York: Cambridge University Press, 1998).

14. Quoted in Mark Elvin, "Who Was Responsible for the Weather? Moral Meteorology in Late Imperial China," *Osiris* 13 (1998): 229.

15. Pierre-Etienne Will, *Bureaucracy and Famine in Eighteenth-Century China* (Stanford: Stanford University Press, 1990).

16. James Lee, "The Southwest: Yunnan and Guizhou," in *Nourish the People: The State Civilian Granary System in China*, ed. Pierre-Etienne Will and R. Bin Wong (Ann Arbor: University of Michigan Press, 1991), 465; James Lee, "Food Supply and Population Growth in Southwest China, 1250–1850," *Journal of Asian Studies* 41.4 (August 1982): 741.

17. Will, *Bureaucracy and Famine*, 276.

18. D. Zhang et al., "Volcanic and ENSO Effects in China Simulations and Reconstructions: Tambora Eruption, 1815," *Climate of the Past: Discussions* 7 (2011): 2071–72.

19. A tree-ring study on the Tibetan plateau found "an abnormal cold period from 1816 to 1822": Eryuan Liang et al., "Tree-Ring Based Summer Temperature Reconstruction for the Source Region of the Yangtze River on the Tibetan Plateau," *Global and Planetary Change* 61 (2008): 318. A dendrochronological study in Nepal produced similar results: Edward R. Cook et al., "Dendroclimatic Signals in Long Tree-Ring Chronologies from the Himalayas of Nepal," *International Journal of Climatology* 23 (2003): 707–32.

20. Ma Yao, *Yunnan Jian Shi*, 404.

21. Angus Maddison, *Chinese Economic Performance in the Long Run* (Paris: Development Centre of the Organization for Economic Co-operation and Development, 1998).

22. Giersch, *The Transformation of Qing China's Yunnan Frontier*, 111.

23. Quoted in Jonathan Marshall, "Opium and the Politics of Gangsterism in Nationalist China, 1927–45," *Committee of Concerned Asian Scholars* 8.3 (1976): 19.

24. Francis Nichols, *Through Hidden Shensi* (New York: Scribner's, 1902), 56–57.

CHAPTER SIX
THE POLAR GARDEN

1. Royal Society, *Minutes of Council* 8 (November 20, 1817): 149–53.

2. "The Top of the World: Is the North Pole Turning to Water?" http://www.john-daly.com/polar/arctic.htm; "Historic Variation in Arctic Ice," June 20, 2009, http://www.wattsupwiththat.com/2009/06/20/historic-variation-in-arctic-ice.

3. *Hamlet* V.ii.379–84.

4. John Barrow, "Lord Selkirk and the North-West Company," *Quarterly Review* (October 1816): 153, 168.

5. William Scoresby, *Arctic Whaling Journals, 1817–20*, ed. C. Ian Jackson (London: Hakluyt Society, 2008–9), 2:45–46.

6. Quoted in Tom Stamp and Cordelia Stamp, *William Scoresby: Arctic Scientist* (Whitby: Caedmon Press, 1976), 65.

7. Stommel and Stommel, *Volcano Weather*, 43.

8. Stamp and Stamp, *William Scoresby: Arctic Scientist*, 66.

9. Otto von Kotzebue, *A Voyage of Discovery into the South Seas and Beering's* [*sic*] *Straits for the Purpose of Exploring a North-East Passage in the Years 1815–18* (London, 1821), 1:212–13.

10. Quoted in Fergus Fleming, *Barrow's Boys* (London: Granta, 1998), 35.

11. John Barrow, "On the Polar Ice and Northern Passage into the Pacific," *Quarterly Review* (October 1817): 199–200.

12. Ibid., 206.

13. *Queen Mab*, 8:58–69.

14. Barrow, "On the Polar Ice," 204.

15. Bernard O'Reilly, *Greenland, the Adjacent Seas, and the North-West Passage to the Pacific Ocean, Illustrated in a Voyage to Davis's Strait, during the Summer of 1817* (London, 1818), iii.

16. Ibid., v–vi, 176.

17. Ibid., 129–30.

18. John Barrow, "O'Reilly's Voyage to Davis's Strait," *Quarterly Review* (April 1818): 208–9.

19. O'Reilly, *Greenland*, 207.

20. Barrow, "O'Reilly's Voyage," 213.

21. For the most thorough attempt to reconstruct O'Reilly's career, see Adriana Craciun, "What Is an Explorer?" *Eighteenth-Century Studies* 45.1 (Fall 2011): 29–51. See also J. P. O'Connor, "Bernard O'Reilly—Genius or Rogue?" *Irish Naturalists' Journal* 21.9 (1985): 379–84.

22. William Scoresby, *Arctic Whaling Journals, 1814–16*, ed. C. Ian Jackson (London: Hakluyt Society, 2008–9), 231.

23. Michael Chenoweth, "Ships' Logbooks and 'The Year without a Summer,'" *Bulletin of the American Meteorological Society* (1996): 2089.

24. Kevin E. Trenberth and Aiguo Dai, "Effects of the Mount Pinatubo Volcanic Eruption on the Hydrological Cycle as an Analog of Geoengineering," *Geophysical Research Letters* 34 (2007): L15702; Eliseev and Mokhov, "Influence of Volcanic Activity on Climate Change in the Past Several Centuries."

25. On the impact of major volcanic eruptions on the AMOC, see Georgiy Stenchikov et al., "Volcanic Signals in Oceans," *Journal of Geophysical Research* 114 (2009): D16104; for the impact of an enhanced AMOC on Arctic sea ice, see Salil Mahajan et al., "Impact of the Atlantic Meridional Overturning Circulation (AMOC) on Arctic Surface Air

Temperature and Sea Ice Variability," *Journal of Climate* 24 (December 2011): 6573–81, and Christophe Kinnard et al., "Reconstructed Changes in Arctic Sea Ice over the Past 1,450 Years," *Nature* 479 (November 24, 2011): 10581. For a geoclimatological overview, see Kevin E. Trenberth and David P. Stepaniak, "The Flow of Energy through the Earth's Climate System," *Quarterly Journal of the Royal Meteorological Society* 130 (October 2004): 2677–2701.

26. A.J.W. Catchpole and Marci-Anne Faurer, "Summer Sea Ice Severity in Hudson Strait, 1751–1870," *Climatic Change* 5 (1983): 115–39; and "Ships' Log-Books, Sea Ice and the Cold Summer of 1816 in Hudson Bay and Its Approaches," *Arctic* 38.2 (June 1985): 121–28.

27. Eleanor Porden, "The Arctic Expeditions: A Poem" (London, 1818), ll. 186–87.

28. More than a century and a half since his death, additions to the vast library on Franklin are made on a nearly annual basis. Recent titles include Martin Sandler, *Resolute: The Epic Search for the Northwest Passage and John Franklin, and the Discovery of the Queen's Ghostship* (New York: Sterling, 2006); Anthony Brandt, *The Man Who Ate His Boots: The Tragic History of the Search for the Northwest Passage* (New York: Knopf, 2010); and Andrew Lambert, *The Gates of Hell: Sir John Franklin's Tragic Quest for the Northwest Passage* (New Haven: Yale University Press, 2011).

29. Dorothy Harley Eber, *Encounters on the Passage: Inuit Meet the Explorers* (Toronto: University of Toronto Press, 2008), 21.

30. On dating Shelley's addition of a polar frame narrative to the *Frankenstein* text, see Charles E. Robinson, *The Frankenstein Notebooks* (New York: Garland, 1996), xxv–vi.

31. Barrow, "Lord Selkirk," 164; for analyses of *Frankenstein* in the context of renewed polar exploration in the late 1810s, and the accompanying explosion in publishing on polar themes, see Jessica Richard, "'A Paradise of My Own Creation': *Frankenstein* and the Improbable Romance of Polar Exploration," *Nineteenth-Century Contexts* 25.4 (2003): 295–314; Jen Hill, *White Horizon: The Arctic in the Nineteenth-Century British Imagination* (Albany: SUNY Press, 2008); and Adriana Craciun, "Writing the Disaster: Franklin and *Frankenstein*," *Nineteenth-Century Literature* 65.4 (2011): 433–80.

32. Shelley, *Frankenstein*, 13, 208; Barrow, "Lord Selkirk," 166.

33. Shelley, *Frankenstein*, 207, 206.

34. *Don Juan* (Canto 1:132).

CHAPTER SEVEN
ICE TSUNAMI IN THE ALPS

1. *Morning Chronicle*, London, October 4, 1817.

2. C. R. Harington, ed., *The Year without a Summer? World Climate in 1816* (Ottawa: Canadian Museum of Nature, 1992), 416.

3. *Letters of Percy Bysshe Shelley*, ed. Jones, 1:497.

4. Douglas I. Benn and David J. A. Evans, *Glaciers and Glaciation* (London: Arnold, 1998), 4; Roy M. Koerner, "Mass Balance of Glaciers in the Queen Elizabeth Islands, Nunavut, Canada," *Annals of Glaciology* 42 (2005): 417–23.

5. *Letters of Percy Bysshe Shelley*, ed. Jones, 1:497–98.

6. Louis Simond, *Switzerland; or, a Journal of a Tour and Residence in that Country, in the Years 1817, 1818, and 1819* (London, 1821), 1:259.

7. Christian Pfister, "Little Ice Age–Type Impacts and the Mitigation of Social Vulnerability to Climate in the Swiss Canton of Bern Prior to 1800," in *Sustainability or Collapse? An Integrated History and Future of People on Earth*, ed. Robert Costanza, Lisa J. Graumlich, and Will Steffen (Cambridge, MA: MIT Press, 2007), 196.

8. Jean Grove, *The Little Ice Age* (London: Methuen, 1988), 122.

9. *Letters of Percy Bysshe Shelley*, ed. Jones, 1:499.

10. Emmanuel Le Roy Ladurie, *Times of Feast, Times of Famine: A History of Climate since the Year 1000*, trans. Barbara Bray (New York: Doubleday, 1971), 207–10.

11. *Journals of Mary Shelley*, ed. Paula R. Feldman and Diana Scott-Kilvert (Baltimore: Johns Hopkins University Press, 1987), 117.

12. Shelley, *Frankenstein*, 95.

13. Ignace Mariétan, "La Vie et L'Oeuvre de L'Ingenieur Ignace Venetz, 1788–1859," *Bulletin de la Murithienne: Société Valaisainne des Sciences Naturelles* 73 (1956): 1–51.

14. The fullest contemporary account of the Val de Bagnes flood is given by H. C. Escher, "A Description of the Val de Bagnes in the Bas Valais, and of the Disaster which Befell It in June, 1818," *Blackwood's Edinburgh Magazine* (October 1818–March 1819): 87–95; see also Philippe-Sirice Bridel, *Seconde Course à la Vallée de Bagnes, et Détails sur les Ravages Occasionnés par L'Ecoulement du Lac de Mauvoisin* (Vevey: Loertscher et Fils, 1818).

15. Benn and Evans, *Glaciers and Glaciation*, 117.

16. F. A. Forel, "Jean-Pierre Perraudin de Lourtier," *Bulletin Société Vaudoise des Sciences Naturelles* 35 (1899): 104–13.

17. H. Lebert, "Biographie de Jean de Charpentier," *Actes de la Société Helvetique des Sciences Naturelles* (August 1877): 140–64.

18. Jean de Charpentier, *Essai sur les Glaciers* (Lausanne, 1841), 421.

19. Escher, "A Description of the Val de Bagnes," 91.

20. William Brockedon, *Illustrations of the Alps* (London, 1828), 1:3.

21. Frank F. Cunningham, *James David Forbes: Pioneer Scottish Glaciologist* (Edinburgh: Scottish Academic Press, 1990), 44.

22. Benn and Evans, *Glaciers and Glaciation*, 9.

23. Ignace Venetz, "Mémoire sur les variations de la température dans les Alpes de la Suisse," *Mémoires des Société Helvetique des Sciences Naturelles* 1.2 (1833): 38.

24. The Charpentier paper first appeared in English in 1836, as "Account of One of the Most Important Results of the Investigations of M. Venetz, Regarding the Present and Earlier Condition of the Glaciers of the Canton Vallais," *Edinburgh New Philosophical Journal* 21 (April–October 1836): 210–20.

25. Jean-Paul Schaer, "Agassiz et les Glaciers: Sa Conduite de la Recherche et Ses Merites," *Eclogae Geologicae Helvetique* 93 (2000): 231–56.

CHAPTER EIGHT
THE OTHER IRISH FAMINE

1. *Letters of John Keats*, ed. Hyder Edward Rollins (Cambridge, MA: Harvard University Press, 1958), 1:321–22.

2. William Carleton, *The Life of William Carleton*, ed. David O'Donoghue (London, 1896), 57.

3. William Carleton, *The Black Prophet: A Tale of Irish Famine* (London, 1847), vi.

4. Thomas Flanagan, *The Irish Novelists, 1800–1850* (New York: Columbia University Press, 1959), 321.

5. Carleton, *Black Prophet*, 24, 22.

6. William Harty, *An Historic Sketch of the Causes, Progress and Present State of the Contagious Fever Epidemic in Ireland* (Dublin, 1819), 113–15.

7. William Kidd, "A Concise Account of the Typhus Fever, at Present Prevalent in Ireland, as it Presented Itself to the Author in One of the Towns in the North of that Country," *Edinburgh Medical and Surgical Journal* 14 (1818): 145.

8. Francis Barker and John Cheyne, *An Account of the Rise, Progress, and Decline of the Fever Lately Epidemical in Ireland* (Dublin, 1821), 31.

9. Harington, *The Year without a Summer*, 368.

10. Barker and Cheyne, *An Account of the Rise*, 30; Carleton, *Black Prophet*, 23.

11. Ibid., 247.

12. Howard, *Climate of London*, 2:6.

13. L. A. Clarkson and E. Margaret Crawford, *Feast and Famine: Food and Nutrition in Ireland, 1500–1920* (New York: Oxford University Press, 2001), 87.

14. D. J. Corrigan, *Famine and Fever as Cause and Effect in Ireland* (Dublin, 1846), 17.

15. Carleton, *Black Prophet*, 26.

16. Harington, *The Year without a Summer*, 498.

17. Kidd, "A Concise Account of the Typhus Fever," 146.

18. Francis Rogan, *Observations on the Condition of the Middle and Lower Classes in the North of Ireland, as It Tends to Promote the Diffusion of Contagious Fever, etc.* (London, 1819), 80.

19. Didier Raoult et al., "The History of Epidemic Typhus," *Infectious Diseases Clinics of North America* 18 (2004): 127–40. For general information on lice and typhus, I draw also upon Didier Raoult and Véronique Roux, "The Body Louse as a Vector of Re-emerging Human Diseases," *Clinical Infectious Diseases* 29 (1999): 888–911, and Abdu F. Azad and Charles B. Beard, "Ricksettial Pathogens and Their Arthropod Vectors," *Emerging Infectious Diseases* 4.2 (1998): 179–86.

20. Ewen Kirkness et al., "Genome Sequences of the Human Body Louse and its Primary Endosymbiont Provide Insights into the Permanent Parasitic Lifestyle," *Proceedings of the National Academy of Sciences* 107.27 (July 2010): 12168–73.

21. Raoult et al., "The History of Epidemic Typhus," 137.

22. Carleton, *Black Prophet*, 193.

23. John Trotter, *Walks through Ireland in the Years 1812, 1814, and 1817* (London, 1819), 303.

24. William Parker, *A Plan for the General Improvement of the State of the Poor in Ireland* (Cork, 1816), 2, 8.

25. "Petition from Cork Complaining of the Increase of Poverty: From the Mayor, Sheriffs, & Several Inhabitants," *Hansard*, February 27, 1818.

26. John Gamble, *Views of Society and Manners in the North of Ireland, in a Series of Letters Written in the Year 1818* (London, 1819), 155.

27. Francis Barker, *Medical Report of the House of Recovery and Fever Hospital in Cork Street, Dublin* (Dublin, 1818), 29.

28. Gamble, *Views of Society and Manners*, 169.

29. *Report from the Select Committee on the State of Disease and Condition of the Labouring Poor in Ireland* (London, 1819), 23.

30. Rogan, *Observations on the Condition of the Middle and Lower Classes*, 93.

31. Quoted in Timothy P. O'Neill, "The State, Poverty, and Distress in Ireland, 1815–45" (Ph.D. thesis, University College, Dublin, 1984), 56; see also O'Neill's published article, "Fever and Public Health in Pre-Famine Ireland," *Journal of the Royal Society of Antiquaries of Ireland* 103 (1973): 1–34; Barker and Cheyne, *An Account of the Rise*, 140.

32. Harty, *An Historic Sketch*, vii.

33. *Dublin Evening Post*, October 30, 1817.

34. Barker and Cheyne, *An Account of the Rise*, 63–65.

35. Rogan, *Observations on the Condition of the Middle and Lower Classes*, 76.

36. *Dublin Evening Post*, November 20, 1817.

37. Amartya Sen and Jean Dreze, eds., *The Political Economy of Hunger* (Oxford: Clarendon Press, 1990), 42.

38. Hugh Fenning, "Typhus Epidemic in Ireland, 1817–19: Priests, Ministers, Doctors," *Collectanea Hibernica* 41 (1999): 117–39.

39. Harty, *An Historic Sketch*, 38.

40. Ibid., 44–47, 93.

41. Joseph Robins, *The Miasma: Epidemic and Panic in Nineteenth-Century Ireland* (Dublin: Institute of Public Administration, 1995), 58.

42. Harty, *An Historic Sketch*, 102.

43. *Edinburgh Medical and Surgical Journal* 14 (October 1818): 529–30. This review section of the 1818 journal lists no fewer than eleven treatises on the typhus epidemic, all published that year in England or Scotland.

44. *Hansard*, April 6, 1819.

45. This figure is arrived at by combining the "moderate" estimate of fever deaths, 65,000, with high-end estimations of the excess mortality caused by famine in the dearth period of 1816–17, namely 80,000 (O'Neill, "The State, Poverty, and Distress in Ireland," 61; Clarkson and Crawford *Feast and Famine*, 127).

46. *Hansard*, April 6, 1819.

47. Mary Louise Briscoe, ed., *Thomas Mellon and His Times* (Pittsburgh: University of Pittsburgh Press, 1994), 10.

48. *Hansard*, April 6, 1819.

49. O'Neill, "The State, Poverty, and Distress in Ireland," 6.

50. Robins, *The Miasma*, 39.

51. William Carleton, *Traits and Stories of the Irish Peasantry* (Boston, 1911), 4:364.

52. Carleton, *Black Prophet*, 211.

53. Shelley, *Frankenstein*, 142, 214.

54. See M. W. Flinn, "The Poor Employment Act of 1817," *Economic History Review* 14 (1961): 82–92.

55. O'Neill, "The State, Poverty, and Distress in Ireland," 300.

56. Patrick Webb, "Emergency Relief during Europe's Famine of 1817: Anticipated Responses to Today's Humanitarian Disasters," *Journal of Nutrition* 132 (2002): 2092S–2095S. See also John Post, *The Last Great Subsistence Crisis in the Western World* (Baltimore: Johns Hopkins University Press, 1977), 66–67, 175.

57. See Mike Davis's groundbreaking study of the Indian famines of the 1870s, *Late Victorian Holocausts: El Niño Famines and the Making of the Third World* (New York: Verso, 2001).

CHAPTER NINE
HARD TIMES AT MONTICELLO

1. Willis I. Milham, "The Year 1816—The Causes of Abnormalities," *Monthly Weather Review* 52.12 (December 1924): 563.

2. William R. Baron, "1816 in Perspective: The View from the Northeastern United States," in *The Year without a Summer*, ed. Harington, 124–44.

3. Pomroy Jones, *Annals and Recollections of Oneida County* (Rome, NY, 1851), 79.

4. *Daily National Intelligencer*, May 1, 1816, quoted in C. Edward Skeen, "'The Year without a Summer': An Historical View," *Journal of the Early Republic* 1 (Spring 1981): 61.

5. Joseph B. Hoyt, "The Cold Summer of 1816," *Annals of the Association of American Geographers* 48 (1958): 118–31; Michael Chenoweth, "Daily Synoptic Weather Map Analysis of the New England Cold Wave and Snowstorms of 5 to 11 June, 1816," in *Historical Climate Variability and Impacts in North America*, ed. L. A. Dupigny-Giroux and C. J. Mock (Berlin: Springer, 2009), 107–21.

6. Quoted in David Ludlum, *Early American Winters, 1604–1820* (Boston: American Meteorological Society, 1966), 190.

7. Stommel and Stommel, *Volcano Weather*, 37; Hoyt, "The Cold Summer of 1816," 122.

8. Skeen, "'The Year without a Summer': An Historical View," 56; Stommel and Stommel, *Volcano Weather*, 42, 63.

9. Thomas Jefferson, *Writings*, ed. Paul Leicester Ford (New York: G. P. Putnam's Sons, 1892–99), 10:64–65, 52.

10. Thomas Jefferson, *Farm Book*, ed. Edwin Morris Betts (Princeton: Princeton University Press, 1953), 221.

11. Jefferson, *Writings*, ed. Ford, 10:52.

12. Georges Louis Leclerc, Comte de Buffon, "Des époques de la nature," *Histoire Naturelle* (Paris, 1778), 5:239.

13. Ibid., 5:168.

14. Georges Louis Leclerc, Comte de Buffon, *Buffon's Natural History, Containing a Theory of the Earth, a General History of Man, of the Brute Creation, and of Vegetables, Minerals, etc.* (London, 1797), 7:43, 27, 40, 38.

15. Ibid., 7:42, 48.

16. On Shelley and Buffon, see Alan Bewell, *Romanticism and Colonial Disease* (Baltimore: Johns Hopkins University Press, 1999), 212–14.

17. *Queen Mab*, 8:145–54.

18. The most thorough scholarly account is by Italian historian Antonello Gerbi, *The Dispute of the New World: The History of a Polemic, 1750–1900*, trans. Jeremy Mole (Pittsburgh: University of Pittsburgh Press, 1973). See also James R. Fleming, *Historical Perspectives on Climate Change* (New York: Oxford University Press, 1998).

19. Thomas Jefferson, *Notes on the State of Virginia*, ed. William Peden (Chapel Hill: University of North Carolina Press, 1982), 43–44.

20. Ibid., 60.

21. Thomas Jefferson, *Garden Book, 1766–1824*, ed. Edwin Morris Betts (Philadelphia: American Philosophical Society, 1944), 255; Robert Rogers, *A Concise Account of North America* (London, 1765), 194, 199.

22. Daniel Webster, *Private Correspondence*, ed. Fletcher Webster (Boston: Little, Brown, 1857), 1:371.

23. Buffon, *Buffon's Natural History*, 10:48.

24. In an 1862 lecture, Lord Kelvin reiterated the still-prevailing scientific consensus that "the earth is becoming on the whole cooler from age to age." "On the Secular Cooling of the Earth," *Transactions of the Royal Society of Edinburgh* 23 (1864): 167–69.

25. Buffon, "Des époques," 244.

26. Webster, *Private Correspondence*, 1:371.

27. Buffon, "Des époques," 248.

28. Constantin-François Volney, *View of the Climate and Soil of the United States of America* (London, 1804), 215–16.

29. Samuel Williams, *Natural and Civil History of Vermont*, 2nd ed. (Burlington, 1809), 70–71.

30. Hoyt, "The Cold Summer of 1816," 126.

31. Quoted in Post, *The Last Great Subsistence Crisis in the Western World*, 106.

32. Thomas Jefferson, *Writings*, ed. Andrew Lipscomb (Washington, DC: Thomas Jefferson Memorial Association, 1903–4), 15:141.

33. See P. D. Jones et al., "Tree-Ring Evidence of the Widespread Effects of Explosive Volcanic Eruptions," *Geophysical Research Letters* 22.11 (June 1995): 1333–36; and K. R. Briffa et al., "Tree-Ring Reconstructions of Summer Temperature Patterns across Western North America since 1600," *Journal of Climate* 5 (1992): 735–54.

34. Leland D. Baldwin, *The River Boat on Western Waters* (Pittsburgh: University of Pittsburgh Press, 1941), 181; *Niles' Weekly Register*, July 11, 1818.

35. I owe a great debt to the Illinois historian David M. Brady for wide-ranging conversations and research assistance exploring the teleconnections between the Tamboran volcanic weather of 1816–18 in the United States and the subsequent financial crisis. His essay "The Panic of 1819: Its Cause and Effects in Illinois History," (Springfield: Sangamon County Historical Society, 2006), and his unpublished text, "Geological and Economic Influences on Society: The Tambora Eruption of 1815 and the Panic of 1819, a Case Study in Illinois History," were important sources for this chapter. John Post, in the final pages of *The Last Great Subsistence Crisis in the Western World*, was the first to speculate on the relationship between the post-Tambora agricultural recovery of 1818 in Europe and the Panic of 1819 in the United States.

36. Quoted in Hoyt, "The Cold Summer of 1816," 126.

37. Clarence Albert Day, *A History of Maine Agriculture, 1604–1860* (Orono: Maine University Press, 1954), 111–13.

38. Morris Birkbeck, *Notes on a Journey in America, from the Coast of Virginia to the Territory of Illinois, with Proposals for the Establishment of a Colony of English* (Ann Arbor: University of Michigan Press, 1968), 30; quoted in Malcolm J. Rohrbough, *The Land Office Business: The Settlement and Administration of American Public Lands, 1789–1857* (New York: Oxford University Press, 1968), 108; Jefferson, *Writings*, ed. Ford, 10:115.

39. Day, *A History of Maine Agriculture*, 111.

40. Henry Fearon, *Sketches of America: A Narrative of a Journey of Five Thousand Miles through the Eastern and Western States of America* (London, 1818), vii.

41. Adam Seybert, *Statistical Annals of the United States, 1789–1818* (Philadelphia, 1818), 28–29.

42. H.J.M. Johnston, *British Emigration Policy, 1815–1830: "Shovelling Out Paupers"* (Oxford: Clarendon Press, 1972), 29.

43. Quoted in R. Carlyle Buley, *The Old Northwest: Pioneer Period, 1815–40* (Bloomington: Indiana University Press, 1950), 12, 6; Rohrbough, *The Land Office Business*, 109.

44. Henry Schoolcraft, *Personal Memoirs of a Residence of Thirty Years with the Indian Tribes on the American Frontiers, with Brief Notices of Passing Events, Facts, and Opinions, AD 1812 to AD 1842* (Philadelphia, 1851), 19.

45. Scott Derks and Tony Smith, eds., *The Value of a Dollar, 1600–1865* (Milberton, NY: Grey House Publishers, 2005), 232; quoted in Daniel S. Dupre, "The Panic of 1819 and the Political Economy of Sectionalism," in *The Economy of Early America: Historical Perspectives and New Directions*, ed. Cathy Matson (University Park: Pennsylvania State University Press, 2006), 277.

46. Dumas Malone, *Jefferson and His Time*, 6 vols. (Boston: Little, Brown, 1948–81), 6:310.

47. Clyde Haulman, *Virginia and the Panic of 1819: The First Great Depression and the Commonwealth* (London: Pickering & Chatto, 2008), 11; William Darby, *The Emigrant's Guide to the Western and Southwestern States and Territories* (New York, 1818), 46.

48. Derks and Smith, *Value of a Dollar*, 245, 262.

49. Malone, *Jefferson and His Time*, 6:309.

50. Jefferson, *Writings*, ed. Ford, 10:134.

51. Jefferson, *Writings*, ed. Lipscomb, 15:279.

52. Jefferson, *Writings*, ed. Ford, 10:157.

53. Daniel S. Dupre, *Transforming the Cotton Frontier: Madison County, Alabama, 1800–40* (Baton Rouge: Louisiana State University Press, 1997), 58.

54. Haulman, *Virginia and the Panic of 1819*, 32; Hattie M. Anderson, "Frontier Economic Problems in Missouri, 1815–28," *Missouri Historical Review* 34.1 (1939): 38–70; Dorothy D. Dorsey, "The Panic of 1819 in Missouri," *Missouri Historical Review* 29.2 (1935): 79–91.

55. Quoted in Samuel Rezneck, "The Depression of 1819–22: A Social History," *American Historical Review* 39.1 (1933): 30; Dupre, *Transforming the Cotton Frontier*, 55; Haulman, *Virginia and the Panic of 1819*, 25.

56. James Flint, *Letters from America, Containing Observations on the Climate and Agriculture of the Western States, etc.* (Edinburgh, 1822), 200; quoted in Andrew R. L. Cayton, "The Fragmentation of 'A Great Family': The Panic of 1819 and the Rise of the Middling Interest in Boston, 1818–22," *Journal of the Early Republic* 2 (Summer 1982): 146.

57. Haulman, *Virginia and the Panic of 1819*, 50; Leon Schur, "The Second Bank of the United States and the Inflation after the War of 1812," *Journal of Political Economy* 68.2 (April 1960): 132.

58. Dupre, "The Panic of 1819," 277.

59. See Robert V. Remini, *Andrew Jackson and the Bank War: A Study in the Growth of Presidential Power* (New York: Norton, 1967).

60. David Hackett Fischer, *The Great Wave: Price Revolutions and the Rhythm of History* (New York: Oxford University Press, 1996), 155.

61. Edgar E. Hume, "The Foundation of American Meteorology by the United States Army Medical Department," *Bulletin of the History of Medicine* 8 (1940): 205–8.

62. Edward J. Hopkins and Joseph M. Moran, "Monitoring the Climate of the Old Northwest, 1820–95," in *Historical Climate Variability and Impacts*, 174, 180, 186.

63. The Dustbowl of the 1930s figures as an exception that proves the rule, hence its notoriety.

EPILOGUE
ET IN EXTREMIS EGO

1. Hans-Erhard Lessing, "What Led to the Invention of the Early Bicycle?" *Cycle History* 11 (2001): 28–36.

2. Paul Crutzen, "Albedo Enhancement by Stratospheric Sulfur Injections: A Contribution to Resolve a Policy Dilemma?" *Climatic Change* 77 (2006): 211–19: for other significant contributions to the debate, see Alan Robock et al., "The Benefits, Risks, and Costs of Stratospheric Engineering," *Geophysical Research Letters* (2009): L19703; and J. C. Moore et al., "Efficacy of Geoengineering to Limit 21st Century Sea-Level Rise," *Proceedings of the National Academy of Sciences* 107.36 (September 2010): 15699–703. For a history of twentieth-century geo-engineering research and politics, see James R. Fleming, *Fixing the Sky: The Checkered History of Weather and Climate Control* (New York: Columbia University Press, 2010).

3. See J.-P. Vernier et al., "Major Influence of Tropical Volcanic Eruptions on the Stratospheric Aerosol Layer during the Last Decade," *Geophysical Research Letters* 38 (June 2011): L12807; and R. R Neely et al., "Recent Anthropogenic Increases in SO_2 from Asia Have Minimal Impact on Stratospheric Aerosol," *Geophysical Research Letters* (published online, March 2013).

4. Biologist Eugene Stoermer first coined the term "Anthropocene" in the 1980s. See the *International Geosphere-Biosphere Programme Newsletter* 41 (2000) for its first published definition.

5. The scientific documentation of current Arctic sea-ice decline is vast and demoralizing. One recent study, based on analysis of marine sediments west of the island of Svalbard (Arctic waters well-known to William Scoresby), concluded the current high polar ocean temperatures to be unique in at least the last two thousand years and thus "linked to the Arctic amplification of global warming": Robert F. Spielhagen et al., "Enhanced Modern Heat Transfer to the Arctic by Warm Atlantic Water," *Science* 331 (January 2011): 450–53. Current warming of the Arctic thus relates to the same proximate physical causes as those described in chapter 6, only greatly enhanced.

6. Shelley, *Frankenstein*, 96.

GENERAL BIBLIOGRAPHY

I have relied on two standard references texts on volcanology and climate, respectively: Peter Francis and Clive Oppenheimer, *Volcanoes*, 2nd ed. (Oxford: Oxford University Press, 2004) and Dennis L. Hartmann, *Global Physical Climatology* (San Diego: Academic Press, 1994). Where scientific information is not otherwise cited, I have drawn it from these sources.

MOUNT TAMBORA, VOLCANISM, AND CLIMATE
(WORKS CONSULTED BUT NOT CITED)

Anchukaitis, K. J., et al. "Influence of Volcanic Eruptions on the Climate of the Asian Monsoon Region." *Geophysical Research Letters* 37 (2010): L22703.

Crowley, Thomas J. "Causes of Climate Change Over the Past 1000 Years." *Science* 289 (July 2000): 270–77.

De Angelis, M., et al. "Volcanic Eruptions Recorded in the Illimani Ice Core (Bolivia): 1918–1998 and Tambora Periods." *Atmospheric Chemistry and Physics* 3 (2003): 1725–41.

De Boer, Jelle Zelinga, and Donald T. Sanders. *Volcanoes in Human History: The Far-Reaching Effects of Major Eruptions*. Princeton: Princeton University Press, 2002.

Gao, Chaochao, et al. "Atmospheric Volcanic Loading Derived from Bipolar Ice Cores: Accounting for the Spatial Distribution of Volcanic Deposition." *Journal of Geophysical Research* 112 (2007): D09109.

Haigh, Joanna D., et al. "The Response of Tropospheric Circulation to Perturbations in Lower-Stratospheric Temperature." *Journal of Climate* 18 (2005): 3672–85.

Hameed, Sultan, et al. "Climate in China after the Tambora Eruption." *Carbon Dioxide Information Analysis Center (CDIAC) Communications* (1989): 6–8.

Klingaman, William K., and Nicholas P. Klingaman. *The Year without Summer: 1816 and the Volcano That Darkened the World and Changed History*. New York: St. Martin's, 2013.

Lamb, Hubert H. *Climate: Present, Past, and Future*. 2 vols. London: Metheun, 1972–77.

———. *Weather, Climate, and Human Affairs*. London: Routledge, 1988.

Jones, P. D., et al., eds. *History and Climate: Memories of the Future?* New York: Kluwer Academic/Plenum Publishers, 2001.

McCormick, M. Patrick, et al. "Atmospheric Effects of the Mt. Pinatubo Eruption." *Nature* 373 (February 2, 1995): 399–404.

Pisek, Jan, and Rudolf Brazdil. "Responses of Large Volcanic Eruptions in the Instrumental and Documentary Climatic Data over Central Europe." *International Journal of Climatology* 26 (2006): 439–59.

Rampino, Michael R. "Distant Effects of the Tambora Eruption of April, 1815." *Eos* 70.51 (December 1989): 1559–60.

———. "Historic Eruptions of Tambora (1815), Krakatau (1883), and Agung (1963), Their Stratospheric Aerosols, and Climatic Impact." *Quaternary Research* 18 (1982): 127–43.

Robertson, A., et al. "Hypothesized Climate Forcing Time Series for the Last 500 Years." *Journal of Geophysical Research* 106 (July 2001): 14,783–803.

Rotbert, Robert I., and Theodore K. Rabb, eds. *Climate and History*. Princeton: Princeton University Press, 1981.

Self, S. "The Effects and Consequences of Very Large Explosive Volcanic Eruptions." *Philosophical Transactions of the Royal Society* 364 (2006): 2073–97.

Shindell, Drew T., and Gavin Schmidt. "Volcanic and Solar Forcing of Climate Change during the Preindustrial Era." *Journal of Climate* 16 (2003): 4094–4107.

Sigurdsson, Haraldur. "Evidence of Volcanic Loading of the Atmosphere and Climate Response." *Palaeogeography, Palaeoclimatology, Palaeoecology* 89 (1990): 277–89.

Stommel, Henry, and Elizabeth Stommel. *Volcano Weather: The Story of 1816, the Year without a Summer*. Newport, RI: Seven Seas Press, 1983.

Trigo, Ricardo M., et al. "Iberia in 1816, the Year without a Summer." *International Journal of Climatology* 29 (2009): 99–115.

Vupputuri, R.K.R. "The Tambora Eruption in 1815 Provides a Test on Possible Global Climatic and Chemical Perturbations in the Past." *Natural Hazards* 5 (1992): 1–16.

Weeckstee, O., and G. Weeckstee. "L'Eruption du Volcan Tambora en 1815: A-t-Elle Eu des Repercussions Climatiques en France?" *Lave* 83 (February 2000): 19–25.

Wigley, T.M.L., M. J. Ingram, and G. Farmer, eds. *Climate and History: Studies in Past Climates and Their Impact on Man*. Cambridge: Cambridge University Press, 1981.

Zeilinga de Boer, Jelle, and Donald Theodore Sanders. *Volcanoes in Human History: The Far-Reaching Effects of Major Eruptions*. Princeton: Princeton University Press, 2002.

Zollinger, Heinrich. *Besteigung des Vulkanes Tambora auf der Insel Sumbawa und Schilderung der Erupzion im Jahr 1815*. Winterthur, 1855.

MARY SHELLEY AND HER CIRCLE

Byron, Lord. *The Complete Poetical Works*. Ed. Jerome J. McGann. 7 vols. Oxford University Press, 1980–93.

———. *Letters and Journals*. Ed. Leslie A. Marchand. 12 vols. Cambridge, MA: Harvard University Press, 1973–82.

Ellis, David. *Byron in Geneva: That Summer of 1816*. Liverpool: Liverpool University Press, 2011.

Holmes, Richard. *Shelley: The Pursuit*. New York: E. P. Dutton & Co., 1975.

Keats, John. *Complete Poems*. Ed. Jack Stillinger. Cambridge, MA: Harvard University Press, 1978.

———. *Letters*. Ed. Hyder Edward Rollins. 2 vols. Cambridge, MA: Harvard University Press, 1958.

Lovell, Ernest J. *Captain Medwin: Friend of Byron and Shelley*. Austin: University of Texas Press, 1962.

———, ed. *His Very Self and Voice: Collected Conversations of Lord Byron*. New York: Macmillan, 1954.

Medwin, Thomas. *The Angler in Wales; or, Days and Nights of Sportsmen*. 2 vols. London, 1834.

Polidori, John William. *Diary 1816: Relating to Byron, Shelley, etc*. Ed. William Michael Rossetti. London, 1911.

Robinson, Charles E. *The Frankenstein Notebooks*. New York: Garland, 1996.

Seymour, Miranda. *Mary Shelley*. London: John Murray, 2000.

Shelley, Mary. *Frankenstein; or, The Modern Prometheus*. Ed. Maurice Hindle. New York: Penguin, 1992.

———. *Journals*. Ed. Paula R. Feldman and Diana Scott-Kilvert. Baltimore: Johns Hopkins University Press, 1987.

———. *The Last Man*. Oxford: Oxford University Press, 2008.

———. *Letters*. Ed. Betty T. Bennett. 3 vols. Baltimore: Johns Hopkins University Press, 1980–88.

Shelley, Percy Bysshe. *Complete Poetical Works*. Ed. Neville Rogers. 2 vols. Oxford: Clarendon Press, 1972–.

———. *Letters*. Ed. Frederick L. Jones. 2 vols. Oxford: Clarendon Press, 1964.

———. *The Prose Works of Percy Bysshe Shelley*. Ed. E. B. Murray. 2 vols. Oxford: Clarendon Press, 1993.

———. *Shelley's Poetry and Prose*. Ed. Donald Reiman and Sharon B. Powers. New York: Norton, 1977.

Stocking, Marion Kingston, ed. *The Clairmont Correspondence: Letters of Claire Clairmont, Charles Clairmont, and Fanny Imlay Godwin*. 2 vols. Baltimore: Johns Hopkins University Press, 1995.

Todd, Janet. *Death and the Maidens: Fanny Wollstonecraft and the Shelley Circle*. Berkeley: Counterpoint Press, 2007.

BIBLIOGRAPHY BY CHAPTER

INTRODUCTION
FRANKENSTEIN'S WEATHER

Franklin, Benjamin. "Meteorological Imaginations and Conjectures." *Memoirs of the Literary and Philosophical Society of Manchester* 2 (1784): 357–61.

Zerefos, C. S., et al. "Atmospheric Effects of Volcanic Eruptions as Seen by Famous Artists and Depicted in Their Paintings." *Atmospheric Chemistry and Physics* 7 (2007): 4027–42.

CHAPTER ONE
THE POMPEII OF THE EAST

Abeyasekere, Susan. "Slaves in Batavia: Insights from a Slave Register." In *Slavery, Bondage, and Dependency in Southeast Asia*, ed. Anthony Reid, 315–40. New York: St. Martin's, 1983.

Assey, Charles. *On the Trade to China, and the Indian Archipelago, with Observations on the Insecurity of British Interests in That Quarter*. London, 1819.

Boers, Bernice de Jong. "The 'Arab' of the Indonesian Archipelago: The Famed Horse Breeds of Sumbawa." In *Breeds of Empire: The "Invention" of the Horse in Southern Africa and Maritime South East Asia, 1500–1950*, ed. Greg Bankoff and Sandra Swart, 51–64. Copenhagen: NIAS Press, 2009.

———. "A Volcanic Eruption in Indonesia and Its Aftermath." *Indonesia* 60 (October 1995): 37–60.

Boomgaard, Peter. *Children of the Colonial State: Population Growth and Economic Development in Java, 1795–1880*. Amsterdam: Free University Press, 1989.

Carey, Peter, ed. *The British in Java, 1811–1816: A Javanese Account*. New York: Oxford University Press, 1992.

Crawfurd, John. *A Descriptive Dictionary of the Indian Islands and Adjacent Countries.* New York: Oxford University Press, 1971.

D'Arrigo, Roseanne, et al. "Monsoon Drought over Java, Indonesia, during the Past Two Centuries." *Geophysical Research Letters* 33 (2006): L04709.

Foden, J. "The Petrology of Tambora Volcano: A Model for the 1815 Eruption." *Journal of Volcanology and Geothermal Research* 27 (1986): 1–41.

Gertisser, Ralf, Stephen Self, et al. "Processes and Timescales of Magma Genesis and Differentiation Leading to the Great Tambora Eruption in 1815." *Journal of Petrology* 53.2 (2012): 271–97.

Goethals, Peter R. *Aspects of Local Government in a Sumbawan Village.* Ithaca: Cornell University Department of Far Eastern Studies, 1961.

Hitchcock, M. J. "Is This Evidence for the Lost Kingdoms of Tambora?" *Indonesia Circle* 33 (1984): 30–35.

Kathirithamby-Wells, Jeyamalar. "Socio-political Structures and the South-East Asian Ecosystem: An Historical Perspective up to the Mid-Nineteenth Century." In *Asian Perceptions of Nature: A Critical Approach*, ed. Ole Brunn and Arne Kalland, 25–46. Richmond: Curzon Press, 1995.

Moor, J. H. *Notices of the Indian Archipelago and Adjacent Countries.* Singapore, 1837.

Munshi, Abdullah. *Autobiography.* Trans. W. G. Shellabear. Singapore, 1918.

Oppenheimer, Clive. *Eruptions That Shook the World.* New York: Cambridge University Press, 2011.

Raffles, Sir Thomas Stamford. *The History of Java.* Intro. John Bastin. 2 vols. New York: Oxford University Press, 1965.

Reid, Anthony, ed. *Slavery, Bondage, and Dependency in Southeast Asia.* New York: St. Martin's, 1983.

Ross, J. T. "Narrative of the Effects of the Eruption from the Tomboro [*sic*] Mountain in the Island of Sumbawa." *Batavian Transactions* 8 (1816): 1–25.

Self, Stephen, et al. "Volcanological Study of the Great Tambora Eruption of 1815." *Geology* 12 (November 1984): 659–63.

Self, Stephen, Ralf Gertisser, et al. "Magma Volume, Volatile Emissions, and Stratospheric Aerosols from the 1815 Eruption of Tambora." *Geophysical Research Letters* 31 (2004): L20608.

Sigurdsson, Haraldur, and Steven Carey. "Eruptive History of Tambora Volcano." *Mitteilungen Geologisch-Palaeontologische Institut* (University of Hamburg) 70 (1992): 187–206.

———. "Plinian and Co-ignimbrite Tephra Fall from the 1815 Eruption of Tambora Volcano." *Bulletin of Volcanology* 51 (1989): 243–70

Smith, Adam. *The Wealth of Nations.* New York: Modern Library, 1937.

Stothers, Richard B. "The Great Tambora Eruption in 1815 and Its Aftermath." *Science* 224 (June 15, 1984): 1191–98.

van der Kraan, A. "Bali, Slavery and the Slave Trade." *Slavery, Bondage, and Dependency in South East Asia*, ed. Anthony Reid, 315-40. New York: St. Martin's, 1983.

Warren, James F. *Iranun and Balangingi: Globalization, Maritime Raiding and the Birth of Ethnicity*. Singapore: Singapore University Press, 2002.

———. "A Tale of Two Centuries: The Globalization of Maritime Raiding and Piracy in South East Asia at the End of the Eighteenth and Twentieth Centuries." In *A World of Water: Rain, Rivers, and Seas in South-East Asian Histories*, ed. Peter Boomgaard, 125–52. Leiden: KITLV Press, 2001.

Wood, Gillen D'Arcy. "The Volcano Lover: Climate, Colonialism, and Slavery in Raffles's *History of Java.*" *Journal of Early Modern Cultural Studies* 8.2 (2008): 33–54.

Wright, H.R.C. "Raffles and the Slave Trade at Batavia in 1812." *Historical Journal* 3.2 (1960): 184–91.

Wurtzburg, C. E. *Raffles of the Eastern Isles*. 2nd ed. New York: Oxford University Press, 1984.

Zamoyski, Adam. *Rites of Peace: The Fall of Napoleon and the Congress of Vienna*. New York: Harper Collins, 2007.

CHAPTER TWO
THE LITTLE (VOLCANIC) ICE AGE

Alley, Richard B. *The Two-Mile Time Machine: Ice Cores, Abrupt Climate Change, and Our Future*. Princeton: Princeton University Press, 2000.

Altick, Richard. *The Shows of London*. Cambridge, MA: Harvard University Press, 1978.

Briffa, K. R., et al. "Influence of Volcanic Eruptions on Northern Hemisphere Summer Temperature over the Past 600 Years." *Nature* 393 (June 1998): 450–55.

Cole-Dai, Jihong [Dai, Jihong], et al. "Cold Decade (AD 1810–19) Caused by Tambora (1815) and Another (1809) Stratospheric Volcanic Eruption." *Geophysical Research Letters* 36 (2009): L22703.

Coleridge, Samuel Taylor. "Fancy in Nubibus; or, the Poet in the Clouds." *Blackwood's Magazine* (November 1819).

Crowley, Thomas J., et al. "Volcanism and the Little Ice Age." *PAGES News* 16.2 (April 2008): 22–23.

Dai, Jihong [Cole-Dai, Jihong], et al. "Ice Core Evidence for an Explosive Tropical Volcanic Eruption 6 Years Preceding Tambora." *Journal of Geophysical Research* 96 (September 1991): 17361–66.

Daly, Nicholas. "The Volcanic Disaster Narrative: From Pleasure Garden to Canvas, Page, and Stage." *Victorian Studies* 53.2 (Winter 2011): 255–85.

D'Arrigo, Roseanne, et al. "The Impact of Volcanic Forcing on Tropical Temperatures during the Past Four Centuries." *Nature Geoscience* 2 (2009): 51–56.

Davy, Humphry. "On the Phenomena of Volcanoes." In *Collected Works*, ed. John Davy, 6:344–58. London, 1839–40.

Eliseev, A. V., and I. I. Mokhov. "Influence of Volcanic Activity on Climate Change in the Past Several Centuries: Assessments with a Climate Model of Intermediate Complexity." *Izvestiya: Atmospheric and Oceanic Physics* 44.6 (2008): 671–83.

Foden, J. "The Petrology of Tambora Volcano: A Model for the 1815 Eruption." *Journal of Volcanology and Geothermal Research* 27 (1986): 1–41.

Gao, Chaochao, et al. "Atmospheric Volcanic Loading Derived from Bipolar Ice Cores: Accounting for the Spatial Distribution of Volcanic Deposition." *Journal of Geophysical Research* 112 (2007): D09109.

———. "Volcanic Forcing of Climate over the Past 1500 Years: An Improved Ice Core–Based Index for Climate Models." *Journal of Geophysical Research* 113 (2008): D23111.

Gertisser, Ralf, Stephen Self, et al. "Processes and Timescales of Magma Genesis and Differentiation Leading to the Great Tambora Eruption in 1815." *Journal of Petrology* 53.2 (2012): 271–97.

Hegerl, G., et al. "Detection of Volcanic, Solar, and Greenhouse Signals in Paleo-Reconstructions of Northern Hemisphere Temperature." *Geophysical Research Letters* 30.5 (2003): 1242.

Howard, Luke. *The Climate of London, Deduced from Meteorological Observations, Made at Different Places, in the Neighbourhood of the Metropolis*. 2 vols. London, 1820.

Humboldt, Alexander von. *Personal Narrative of Travels to the Equinoctial Regions of the New Continent during the Years 1799–1804*. Trans. Helen Maria Williams. London, 1815.

Johnston, Arch, and N. K. Moran. "John Weisman Accounts of the Earthquakes." Memphis: Center for Earthquake Research and Information. http://www.ceri.memphis .edu/compendium/eyewitness/wiseman.html.

Mackenzie, George. *The System of the Weather of the British Islands*. Edinburgh, 1821.

Mackenzie, George Steuart. *Travels in the Island of Iceland, during the Summer of 1810*. London, 1811.

Metternich, [Prince]. *Memoirs, 1773–1815*. Ed. Prince Richard Metternich. Trans. Mrs. Alexander Napier. New York, 1881.

Miller, Gifford H. "Abrupt Onset of the Little Ice Age Triggered by Volcanism and Sustained by Sea Ice/Ocean Feedbacks." *Geophysical Research Letters* 39 (January 2012): L02708.

Miller, Mary Ashburn. "Mountain, Become a Volcano: The Image of the Volcano in the Rhetoric of the French Revolution." *French Historical Studies* 32.4 (Fall 2009): 555–85.

Mosley-Thompson, Ellen, et al. "High Resolution Ice Core Records of Late Holocene Volcanism: Current and Future Contributions from the Greenland PARCA Cores." *Volcanism and the Earth's Atmosphere* 139 (2003): 153–64.

Newhall, Christopher G., and Stephen Self. "The Volcanic Explosivity Index (VEI): An Estimate of Explosive Magnitude for Historical Volcanism." *Journal of Geophysical Research* 87 (February 1982): 1231–38.

Oppenheimer, Clive. "Climatic, Environmental, and Human Consequences of the Largest Known Historic Eruption: Tambora Volcano (Indonesia) 1815." *Progress in Physical Geography* 27.2 (2003): 230–59.

———. *Eruptions That Shook the World*. New York: Cambridge University Press, 2011.

———. "Ice Core and Palaeoclimatic Evidence for the Timing and Nature of the Great Mid-13th Century Volcanic Eruption." *International Journal of Climatology* 23 (2003): 417–26.

Palmer, Anne S., et al. "High-Precision Dating of Volcanic Events (AD 1301–1995) Using Ice Cores from Law Dome, Antarctica." *Journal of Geophysical Research* 106 (November 2001): 28089–95.

Robock, Alan. "Cooling Following Large Volcanic Eruptions Corrected for the Effect of Diffuse Radiation on Tree Rings." *Geophysical Research Letters* 32 (2005): L06702.

———. "The 'Little Ice Age': Northern Hemisphere Average Observations and Model Calculations." *Science* 206 (December 1979): 1402–4.

Robock, Alan, and Melissa P. Free. "Ice Cores as an Index of Global Volcanism from 1850 to the Present." *Journal of Geophysical Research* 100 (June 1995): 11549–67.

Robock, Alan, and Yuhe Liu. "The Volcanic Signal in Goddard Institute for Space Studies Three-Dimensional Model Simulations." *Journal of Climate* 7 (1994): 44–55.

Self, Stephen, et al. "Volcanological Study of the Great Tambora Eruption of 1815." *Geology* 12 (November 1984): 659–63.

Siebert, Lee, Tom Simkin, and Paul Kimberley. *Volcanoes of the World*. 3rd ed. Washington, DC: Smithsonian Institution, 2010.

Sigurdsson Haraldur, and Steven Carey. "Eruptive History of Tambora Volcano." *Mitteilungen Geologisch-Palaeontologische Institut* (University of Hamburg) 70 (1992): 187–206.

———. "Plinian and Co-ignimbrite Tephra Fall from the 1815 Eruption of Tambora Volcano." *Bulletin of Volcanology* 51 (1989): 243–70.

Sivertsen, Barbara J. *The Parting of the Sea: How Volcanoes, Earthquakes, and Plagues Shaped the Story of the Exodus*. Princeton: Princeton University Press, 2009.

Staël, Madame de. *Corinne; or, Italy*. Trans. Emily Baldwin and Paulina Driver. London, 1906.

Stendel, Martin, et al. "Influence of Various Forcings on Global Climate in Historical Times Using a Coupled Atmosphere-Ocean General Circulation Model." *Climate Dynamics* 26 (2006): 1–15.

Storch, H. von, et al. "Reconstructing Past Climate from Noisy Data." *Science* 306 (2004): 679–82.

Stothers, Richard B. "Climatic and Demographic Consequences of the Massive Volcanic Eruption of 1258." *Climatic Change* 45 (2000): 361–74.

———. "The Great Tambora Eruption in 1815 and Its Aftermath." *Science* 224 (June 15, 1984): 1191–98.

Timmreck, Claudia, et al. "Limited Temperature Response to the Very Large AD 1258 Volcanic Eruption." *Geophysical Research Letters* 36 (2009): L21708.

Wagner, Sebastian, and Eduardo Zorita. "The Influence of Volcanic, Solar, and CO_2 Forcing on the Temperatures in the Dalton Minimum (1790–1830): A Model Study." *Climate Dynamics* 225 (2005): 205–18.

Wood, Gillen D'Arcy. "Clouds, Constable, Climate Change." *Wordsworth Circle* 38.1–2 (2007): 25–34.

Yalcin, Kaplan, et al. "Ice Core Evidence for a Second Volcanic Eruption around 1809 in the Northern Hemisphere." *Geophysical Research Letters* 33 (2006): L14706.

CHAPTER THREE
"THIS END OF THE WORLD WEATHER"

Brandes, Heinrich. *Beiträgen zur Witterungskunde*. Leipzig, 1820.

Bromwell, William J. *History of Immigration to the United States [1819–55] . . . with an Introductory Review of the Progress and Extent of Immigration to the United States prior to 1819*. New York, 1856.

Clausewitz, Carl von. *Politische Schriften und Briefe*. Ed. Hans Rothfels. Munich: Drei Masken Verlag, 1992.

Clubbe, John. "The Tempest-toss'd Summer of 1816: Mary Shelley's *Frankenstein*." In *Literature in Context*, ed. Joachim Schwend et al. Frankfurt: Peter Lang, 1992.

Coleridge, Samuel Taylor. *Collected Letters*. Ed. Earl Leslie Griggs. 6 vols. Oxford: Clarendon Press, 2000.

Dawson, Alastair, et al. "A 200-Year Record of Gale Frequency, Edinburgh, Scotland: Possible Link with High-Magnitude Volcanic Eruptions." *The Holocene* 7.3 (1997): 337–41.

Fischer, E. M., et al. "European Climate Response to Tropical Volcanic Eruptions over the Last Half Millennium." *Geophysical Research Letters* 34 (2007): L05707.

Forster, Thomas. *Researches into Atmospheric Phaenomena*. London, 1813.

Hamblyn, Richard. *The Invention of Clouds: How an Amateur Meteorologist Forged the Language of the Skies*. New York: Farrar, Straus, and Giroux, 2001.

Henchoz, Paul. "L'Année de la Misère (1816–17) dans la Région Montreux." *Revue Historique Vaudoise* 42 (1934): 66–89.

Henrioud, Marc. "L'Année de la Misère en Suisse et Plus Particulièrement dans le Canton de Vaud." *Revue Historique Vaudoise* 25 (1917): 114–24, 133–42, 171–92.

Howard, Luke. *The Climate of London: Deduced from Meteorological Observations, Made at Different Places, in the Neighbourhood of the Metropolis*. 2 vols. London, 1820.

Lamb, H. H., and A. I. Johnston. *Secular Variations of the Atmospheric Circulation since 1750*. London: Her Majesty's Stationery Office, 1966.

Mackenzie, George. *The System of the Weather in the British Islands*. Edinburgh, 1821.

Masters, Jeff. "2010–2011: Earth's Most Extreme Weather since 1816?" *Weather Underground*. June 24, 2011. www.wunderground.com.

Matthews, Henry. *Diary of an Invalid*. London, 1824.

Peacock, A. J. *Bread or Blood: A Study of the Agrarian Riots in East Anglia in 1816*. London: Victor Gollancz, 1965.

Phillips, Bill. "*Frankenstein* and Mary Shelley's 'Wet, Ungenial Summer.'" *Atlantis* 28.2 (December 2006): 59–68.

Post, John D. *The Last Great Subsistence Crisis in the Western World*. Baltimore: Johns Hopkins University Press, 1977.

Raffles, Thomas. *Letters during a Tour through Some Parts of France, Savoy, Switzerland, Germany, and the Netherlands, in the Summer of 1817*. London, 1818.

Schneider, David P., et al. "Climate Response to Large, High-Latitude and Low-Latitude Volcanic Eruptions in the Community Climate System Model." *Journal of Geophysical Research* 114 (2009): D15101.

Schneider-Carius, Karl. *Weather Science, Weather Research: A History of Their Problems and Findings from Documents during Three Thousand Years.* Trans. from German. New Delhi, 1975.

Shindell, Drew, et al. "Dynamic Winter Response to Large Tropical Volcanic Eruptions since 1600." *Journal of Geophysical Research* 109 (2004): D05104.

Vail, Jeffrey. "'The bright sun was extinguish'd': The Bologna Prophecy and Byron's 'Darkness.'" *Wordsworth Circle* 28.3 (Summer 1997): 183–92.

CHAPTER FOUR
BLUE DEATH IN BENGAL

Arnold, David. *Colonizing the Body: State Medicine and Epidemic Disease in Nineteenth-Century India.* Berkeley: University of California Press, 1993.

Barber, John. *An Account of the Rise and Progress of the Indian or Spasmodic Cholera . . . Illustrated by a Map Showing the Route and Progress of the Disease from Jessore, Near the Ganges, in 1817, to Great Britain, in 1831.* New Haven, 1832.

Blacker, Valentine. *Memoir of the Operations of the British Army in India during the Mahratta War of 1817, 1818, & 1819.* London, 1821.

Burton, R. G. *The Mahratta and Pindari War.* Simla: Government Press, 1910.

Collins, Andrew E. "Vulnerability to Coastal Cholera Ecology." *Social Science and Medicine* 57 (2003): 1397–1407.

Colwell, Rita R. "Emerging and Re-emerging Infectious Diseases: Biocomplexity as an Interdisciplinary Paradigm." *Ecohealth* 2 (2005): 244–57.

———. "Global Climate and Infectious Disease: The Cholera Paradigm." *Science* 274.5295 (December 20, 1996): 2025–31.

———. "Infectious Disease and Environment: Cholera as a Paradigm for Waterborne Disease." *Perspectives: International Microbiology* 7 (2004): 285–89.

Colwell, Rita R., and Anwar Huq. "Marine Ecosystems and Cholera." *Hydrobiologia* 460 (2001): 141–45, 1341–46.

Corbyn, Frederick. *A Treatise on the Epidemic Cholera, as it has Prevailed in India.* Calcutta, 1832.

Cottingham, Kathryn L., et al. "Environmental Microbe and Human Pathogen: The Ecology and Microbiology of *Vibrio Cholerae.*" *Frontiers in Ecology and the Environment* 1.2 (2003): 80–86.

Das, P. K. *The Monsoons: A Perspective.* New Delhi: Indian National Science Academy, 1984.

Ford, Timothy E., et al. "Using Satellite Images of Environmental Changes to Predict Infectious Disease Outbreaks." *Emerging Infectious Diseases* 15.9 (September 2009): 1341–46.

Hamlin, Christopher. *Cholera: The Biography*. New York: Oxford University Press, 2009.

Harrison, Mark. *Climates and Constitutions: Health, Race, Environment and British Imperialism in India, 1600–1850*. New Delhi: Oxford University Press, 1999.

———. *Disease and the Modern World: 1500 to the Present Day*. Cambridge: Polity Press, 2004.

Hastings, Marquess of. *Private Journal*. Ed. Marchioness of Bute. London, 1858.

Heidelberg, J. F. "DNA Sequence of Both Chromosomes of the Cholera Pathogen *Vibrio Cholerae*." *Nature* 406.6795 (August 3, 2000): 477–83.

Higginbotham, J. J. *Selections from the Asiatic Journal and Monthly Register*. Madras, 1875.

Homer. *The Iliad*. Trans. Richmond Lattimore. Chicago: University of Chicago Press, 1951.

Hora, Sunder Lal. "Worship of the Deities, *Olā, Jholū*, and *Bŏn Bībī* in Lower Bengal." *Journal and Proceedings of the Asiatic Society of Bengal* 29 (1933): 1–4.

Jameson, James. *Report on the Epidemick Cholera Morbus, As It Visited the Territories Subject to the Presidency of Bengal in the Years 1817, 1818, and 1819*. Calcutta, 1820.

Jha, Shri Jata Shanka. "The Suppression of the Pindaris." *Journal of the Bihar Research Society* 43 (1954): 251–84.

"Judicial Letters: Measures Adopted for Affording Medical Aid to the Persons Attacked with the Epidemic Disorder at Bewar, Banda and Cawpore." British Library: India Office Records, 1817–18.

Jutla, Antapreet S., et al. "Warming Oceans, Phytoplankton, and River Discharge: Implications for Cholera Outbreaks." *American Journal of Tropical Medicine and Hygiene* 85.2 (2011): 303–8.

Koelle, Katia, et al. "Refractory Periods and Climate Forcing in Cholera Dynamics." *Nature* 436 (August 4, 2005): 696–700.

Lobitz, Brad, et al. "Climate and Infectious Disease: Use of Remote Sensing for Detection of *Vibrio cholerae* by Indirect Measurement." *Proceedings of the National Academy of Science* 97.4 (February 15, 2000): 1438–43.

Moreau de Jonnés, A. *Rapport au Conseil Supérieur de Santé sur le cholera-morbus pestilentiel*. Paris, 1831.

"Papers of George Green Spilsbury." British Library: India Office Records, 1811–27.

Pascual, Mercedes, et al. "Cholera and Climate: Revisiting the Quantitative Evidence." *Microbes and Infection* 4 (2002): 237–45.

———. "Cholera Dynamics and El Niño-Southern Oscillation." *Science* 289 (September 8, 2000): 1766–69.

———. "Hyperinfectivity in Cholera: A New Mechanism for an Old Epidemiological Model?" *Public Library of Science* 3.6 (June 2006): 931–32.

Pascual, Mercedes, and Katia Koelle. "Disentangling Extrinsic from Intrinsic Factors in Disease Dynamics: A Nonlinear Time Series Approach with an Application to Cholera." *American Naturalist* 163.6 (June 2004): 901–13.

Pascual, Mercedes, and Menno Jan Bouma. "Seasonal and Interannual Cycles of Endemic Cholera in Bengal 1891–1940 in Relation to Climate and Geography." *Hydrobiologia* 460 (2001): 147–56.

Prinsep, Henry. *History of the Political and Military Transactions in India during the Administration of the Marquess of Hastings, 1813–23*. London, 1825.

"Proceedings Relative to the Measures which were Successfully Adopted with the View of Counteracting the Fatal Epidemic Denominated Cholera Morbus which Raged in the Town and Suburbs of Calcutta" [1817]. British Library: India Office Records.

Reidl, Joachim, and Karl E. Klose. "*Vibrio cholerae* and Cholera: Out of the Water and into the Host." *FEMS Microbiology Reviews* 26 (2002): 125–39.

Roberts, Emma. *Scenes and Characteristics of Hindostan*. 2 vols. London, 1835.

Robock, Alan. "Volcanic Eruptions and Climate." *Reviews of Geophysics* 38.2 (May 2000): 191–219.

Rodó, Xavier, et al. "ENSO and Cholera: A Nonstationary Link Related to Climate Change?" *Proceedings of the National Academy of Sciences* 99.20 (October 1, 2002): 12901–6.

Schneider, David P., et al. "Climate Response to Large, High-Latitude and Low-Latitude Volcanic Eruptions in the Community Climate System Model." *Journal of Geophysical Research* 114 (2009): D15101.

Shears, Paul. "Recent Developments in Cholera." *Current Opinion in Infectious Diseases* 14 (2001): 553–58.

Statham, James. *Indian Recollections*. London, 1832.

Uma, G., et al. "Recent Advances in Cholera Genetics." *Current Science* 85.11 (December 10, 2003): 1538–45.

Vezzulli, Luigi, et al. "Environmental Reservoirs of *Vibrio cholerae* and Their Role in Cholera." *Environmental Microbiology Reports* 2.1 (2010): 27–33.

von Rad, Ulrich, et al. "A 5000-yr Record of Climate Change in Varved Sediments from the Oxygen Minimum Zone off Pakistan, Northeastern Arabian Sea." *Quarternary Research* 51.1 (1999): 39–51.

Wang, Bin, ed. *The Asian Monsoon*. Berlin: Springer, 2006.

Yadav, Ram R. "Basin Specificity of Climate Change in Western Himalaya, India: Tree-Ring Evidences." *Current Science* 92.10 (May 2007): 1424–29.

Zuckerman, Jane N., et al. "The True Burden and Risk of Cholera: Implications for Prevention and Control." *Lancet: Infectious Diseases* 7 (2007): 521–30.

CHAPTER FIVE
THE SEVEN SORROWS OF YUNNAN

Abel, Clarke. *Narrative of a Journey in the Interior of China . . . in the Years 1816 and 1817*. London, 1818.

Allom, Thomas, and G. N. Wright. *China, in a Series of Views Displaying the Scenery, Architecture, and Social Habits of that Ancient Empire*. 4 vols. London, 1843–47.

Atwill, David. *The Chinese Sultanate: Islam, Ethnicity and the Panthay Rebellion in Southwest China, 1856–73*. Stanford: Stanford University Press, 2005.

Bello, David Anthony. *Opium and the Limits of Empire: Drug Prohibition in the Chinese Interior, 1729–1850*. Cambridge, MA: Harvard University Press, 2005.

Bray, Francesca. *The Rice Economies: Technology and Development in Asian Societies*. London: Blackwell, 1986.

Cao, Shuji. "The Tambora Eruption and Chinese Social History." [Unpublished translation by Fang Wan.] *Academics in China* 5 (September 2009): 37–41.

Chen, Bantian. "The Serious Famine in Yunnan during the Jiaqing Period (1796–1820)." [Unpublished translation by Fang Wan.] *Shi Jie Bo Lan* [*Global Vision*] 8 (2010): 76–79.

Chenglan, Bao, ed. *Synoptic Meteorology in China*. Beijing: China Ocean Press, 1987.

Cook, Edward R., et al. "Dendroclimatic Signals in Long Tree-Ring Chronologies from the Himalayas of Nepal." *International Journal of Climatology* 23 (2003): 707–32.

Deng, Gang. *The Premodern Chinese Economy: Structural Equilibrium and Capitalist Sterility*. New York: Routledge, 1999.

Elvin, Mark. "Who Was Responsible for the Weather? Moral Meteorology in Late Imperial China." *Osiris* 13 (1998): 213–37.

Elvin, Mark, and Liu Ts'uijung. *Sediments of Time: Environment and Society in Chinese History*. New York: Cambridge University Press, 1998.

Giersch, C. Patterson. *The Transformation of Qing China's Yunnan Frontier*. Cambridge, MA: Harvard University Press, 2006.

Lee, James. "Food Supply and Population Growth in Southwest China, 1250–1850." *Journal of Asian Studies* 41.4 (August 1982): 711–46.

———. "The Southwest: Yunnan and Guizhou." In *Nourish the People: The State Civilian Granary System in China*, ed. Pierre-Etienne Will and R. Bin Wong, 431–74. Ann Arbor: University of Michigan Press, 1991.

Lee, Thomas H. C. *Education in Traditional China*. Boston: Brill, 2000.

Liang, Eryuan, et al. "Tree-Ring Based Summer Temperature Reconstruction for the Source Region of the Yangtze River on the Tibetan Plateau." *Global and Planetary Change* 61 (2008): 313–20.

Lin, Mon-Houng. "Late Qing Perceptions of Native Opium." *Harvard Journal of Asiatic Studies* 64.1 (2004): 117–44.

Maddison, Angus. *Chinese Economic Performance in the Long Run*. Paris: Development Centre of the Organization for Economic Co-operation and Development, 1998.

Mariani, Luigi, et al. "Space and Time Behavior of Climatic Hazard of Low Temperature for Single Rice Crop in the Mid Latitude." *International Journal of Climatology* 29 (2009): 1862–71.

Marks, Robert. *Tigers, Rice, Silk, and Silt: Environment and Economy in Late Imperial South China*. New York: Cambridge University Press, 1998.

Marshall, Jonathan. "Opium and the Politics of Gangsterism in Nationalist China, 1927–45." *Committee of Concerned Asian Scholars* 8.3 (1976): 19–43.

Nichols, Francis. *Through Hidden Shensi*. New York: Scribner's, 1902.

Science of the Rice Plant. Tokyo: Food and Agriculture Policy Research Center, 1993–96.

Weather and Rice. Laguna, Philippines: International Rice Research Institute, 1987.

Will, Pierre-Etienne. *Bureaucracy and Famine in Eighteenth-Century China*. Stanford: Stanford University Press, 1990.

Will, Pierre-Etienne, and R. Bin Wong, eds. *Nourish the People: The State Civilian Granary System in China*. Ann Arbor: University of Michigan Press, 1991.

Yang, Bin. *Between Wind and Clouds: The Making of Yunnan*. New York: Columbia University Press, 2009.

Yang, Yuda, Man Zhmin, and Zheng Jinyun, "The Great Yunnan Famine of the Jiaqing Period (1815–17) and the Eruption of the Volcano Mount Tambora." [Unpublished translation by Nicholas Brown.] *Fudan Journal: Social Sciences* (2005): 79–85.

Yao, Ma. *Yunnan Jian Shi* [A Short History of Yunnan]. Kunming: Yunnan ren min chu ban she, 1983.

Yuyang, Li. "Ji yuan shi chao" [Poetry of Li Yuyang]. *Cong Shu Ji Cheng*, vol. 178 (Xinwenfeng chu ban gong si, 1989), 1–103.

Zhang, D., et al. "Volcanic and ENSO Effects in China in Simulations and Reconstructions: Tambora Eruption, 1815." *Climate of the Past: Discussions* 7 (2011): 2061–88.

Zhang, Jiacheng, and Zhiguang Lin. *Climate of China*. Trans. Ding Tan. New York: Wiley & Sons, 1992.

Zhang, Peiyuan. "Extraction of Climate Information from Chinese Historical Writings." *Late Imperial China* 14.2 (December 1993): 96–106.

CHAPTER SIX
THE POLAR GARDEN

Barrow, John. "Burney—Behring's Strait and the Polar Basin." *Quarterly Review* (April 1818): 431–58.

———. "Lord Selkirk and the North-West Company." *Quarterly Review* (October 1816): 129–72.

———. "On the Polar Ice and Northern Passage into the Pacific." *Quarterly Review* (October 1817): 199–223.

———. "O'Reilly's Voyage to Davis's Strait." *Quarterly Review* (April 1818): 208–14.

Brandt, Anthony. *The Man Who Ate His Boots: The Tragic History of the Search for the Northwest Passage*. New York: Knopf, 2010.

Catchpole, A.J.W., and Marci-Anne Faurer. "Ships' Log-Books, Sea Ice and the Cold Summer of 1816 in Hudson Bay and Its Approaches." *Arctic* 38.2 (June 1985): 121–28.

———. "Summer Sea Ice Severity in Hudson Strait, 1751–1870." *Climatic Change* 5 (1983): 115–39.

Chenoweth, Michael. "Ships' Logbooks and 'The Year without a Summer.'" *Bulletin of the American Meteorological Society* (1996): 2077–93.

Craciun, Adriana. "What Is an Explorer?" *Eighteenth-Century Studies* 45.1 (Fall 2011): 29–51.

———. "Writing the Disaster: Franklin and *Frankenstein*." *Nineteenth-Century Literature* 65.4 (2011): 433–80.

Dickson, Bob, and Stephen Dye. "Interrogating the 'Great Ocean Conveyor.' " *Oceanus: Online Magazine of Research from the Woods Hole Oceanographic Institution*, September 6, 2007.

Eber, Dorothy Harley. *Encounters on the Passage: Inuit Meet the Explorers*. Toronto: University of Toronto Press, 2008.

Eliseev, A. V., and I. I. Mokhov. "Influence of Volcanic Activity on Climate Change in the Past Several Centuries: Assessments with a Climate Model of Intermediate Complexity." *Izvestiya: Atmospheric and Oceanic Physics* 44.6 (2008): 671–83.

Fleming, Fergus. *Barrow's Boys*. London: Granta, 1998.

Hill, Jen. *White Horizon: The Arctic in the Nineteenth-Century British Imagination*. Albany: SUNY Press, 2008.

Kinnard, Christophe, et al. "Reconstructed Changes in Arctic Sea Ice over the Past 1,450 Years." *Nature* 479 (November 24, 2011): 10581.

Lambert, Andrew. *The Gates of Hell: Sir John Franklin's Tragic Quest for the Northwest Passage*. New Haven: Yale University Press, 2011.

Mahajan, Salil, et al. "Impact of the Atlantic Meridional Overturning Circulation (AMOC) on Arctic Surface Air Temperature and Sea Ice Variability." *Journal of Climate* 24 (December 2011): 6573–81.

Mills, William James. *Exploring Polar Frontiers: A Historical Encyclopedia*. Santa Barbara: ABC-CLIO, 2003.

O'Connor, J. P. "Bernard O'Reilly—Genius or Rogue?" *Irish Naturalists' Journal* 21.9 (1985): 379–84.

O'Reilly, Bernard. *Greenland, the Adjacent Seas, and the North-West Passage to the Pacific Ocean, Illustrated in a Voyage to Davis's Strait, during the Summer of 1817*. London, 1818.

Porden, Eleanor. "The Arctic Expeditions: A Poem." London, 1818.

Richard, Jessica. " 'A Paradise of My Own Creation': *Frankenstein* and the Improbable Romance of Polar Exploration." *Nineteenth-Century Contexts* 25.4 (2003): 295–314.

Sandler, Martin. *Resolute: The Epic Search for the Northwest Passage and John Franklin, and the Discovery of the Queen's Ghostship*. New York: Sterling, 2006.

Scoresby, William. *An Account of the Arctic Regions*. Edinburgh, 1820.

———. *Arctic Whaling Journals, 1814–16*. Ed. C. Ian Jackson. London: Hakluyt Society, 2008–9.

———. *Arctic Whaling Journals, 1817–20*. Ed. C. Ian Jackson. London: Hakluyt Society, 2008–9.

———. "On the Greenland, or Polar Ice." *Memoirs of the Wernerian Society*. London, 1815.

Stamp, Tom, and Cordelia Stamp. *William Scoresby: Arctic Scientist*. Whitby: Caedmon Press, 1976.

Stenchikov, Georgiy, et al. "Volcanic Signals in Oceans." *Journal of Geophysical Research* 114 (2009): D16104.

Trenberth, Kevin, and Aiguo Dai. "Effects of the Mount Pinatubo Volcanic Eruption on the Hydrological Cycle as an Analog of Geoengineering." *Geophysical Research Letters* 34 (2007): L15702.

Trenberth, Kevin E., and David P. Stepaniak. "The Flow of Energy through the Earth's Climate System." *Quarterly Journal of the Royal Meteorological Society* 130 (October 2004): 2677–2701.

von Kotzebue, Otto. *A Voyage of Discovery into the South Seas and Beering's* [sic] *Straits for the Purpose of Exploring a North-East Passage in the Years 1815–18.* 3 vols. London, 1821.

CHAPTER SEVEN
ICE TSUNAMI IN THE ALPS

Benn, Douglas I., and David J. A. Evans. *Glaciers and Glaciation.* London: Arnold, 1998.

Bridel, Philippe-Sirice. *Seconde Course à la Vallée de Bagnes, et Détails sur les Ravages Occasionnés par L'Ecoulement du Lac de Mauvoisin.* Vevey: Loertscher et Fils, 1818.

Brockedon, William. *Illustrations of the Alps.* London, 1828.

Charpentier, Jean de. "Account of One of the Most Important Results of the Investigations of M. Venetz, Regarding the Present and Earlier Condition of the Glaciers of the Canton Vallais." *Edinburgh New Philosophical Journal* 21 (April–October 1836): 210–20.

———. *Essai sur les Glaciers.* Lausanne, 1841.

Cunningham, Frank F. *James David Forbes: Pioneer Scottish Glaciologist.* Edinburgh: Scottish Academic Press, 1990.

Escher, H. C. "A Description of the Val de Bagnes in the Bas Valais, and of the Disaster which Befell It in June, 1818." *Blackwood's Edinburgh Magazine* (October 1818–March 1819): 87–95.

Forel, F. A. "Jean-Pierre Perraudin de Lourtier." *Bulletin Société Vaudoise des Sciences Naturelles* 35 (1899): 104–13.

Grove, Jean. *The Little Ice Age.* London: Methuen, 1988.

Harington, C. R., ed. *The Year without a Summer? World Climate in 1816.* Ottawa: Canadian Museum of Nature, 1992.

Jamieson, T. F. "On the Ice-Worn Rocks of Scotland." *Quarterly Journal of the Geological Society* 18 (1862): 164–84.

Koerner, Roy M. "Mass Balance of Glaciers in the Queen Elizabeth Islands, Nunavut, Canada." *Annals of Glaciology* 42 (2005): 417–23.

Ladurie, Emmanuel Le Roy. *Times of Feast, Times of Famine: A History of Climate since the Year 1000.* Trans. Barbara Bray. New York: Doubleday, 1971.

Lebert, H. "Biographie de Jean de Charpentier." *Actes de la Société Helvetique des Sciences Naturelles* (August 1877): 140–64.

Mariétan, Ignace. "La Vie et L'Oeuvre de L'Ingenieur Ignace Venetz, 1788–1859." *Bulletin de la Murithienne: Société Valaisainne des Sciences Naturelles* 73 (1956): 1–51.

Pfister, Christian. "Little Ice Age–Type Impacts and the Mitigation of Social Vulnerability to Climate in the Swiss Canton of Bern Prior to 1800." In *Sustainability or Collapse? An Integrated History and Future of People on Earth*, ed. Robert Costanza, Lisa J. Graumlich, and Will Steffen, 197–212. Cambridge, MA: MIT Press, 2007.

Schaer, Jean-Paul. "Agassiz et les Glaciers: Sa Conduite del Recherche et Ses Merites." *Eclogae Geologicae Helvetique* 93 (2000): 231–56.

Simond, Louis. *Switzerland; or, a Journal of a Tour and Residence in that Country, in the Years 1817, 1818, and 1819*. 2 vols. London, 1821.

Venetz, Ignace. "Mémoire sur les variations de la température dans les Alpes de la Suisse." *Mémoires des Société Helvetique des Sciences Naturelles* 1.2 (1833): 1–38.

CHAPTER EIGHT
THE OTHER IRISH FAMINE

Azad, Abu F., and Charles B. Beard. "Rickettsial Pathogens and Their Arthropod Vectors." *Emerging Infectious Diseases* 4.2 (1998): 179–86.

Barker, Francis. *Medical Report of the House of Recovery and Fever Hospital in Cork Street, Dublin*. Dublin, 1818.

Barker, Francis, and John Cheyne. *An Account of the Rise, Progress, and Decline of the Fever Lately Epidemical in Ireland*. Dublin, 1821.

Briscoe, Mary Louise, ed. *Thomas Mellon and His Times*. Pittsburgh: University of Pittsburgh Press, 1994.

Carleton, William. *The Black Prophet: A Tale of Irish Famine*. London, 1847.

———. *The Life of William Carleton*. Ed. David O'Donoghue. London, 1896.

———. *Traits and Stories of the Irish Peasantry*. 4 vols. Boston, 1911.

Clarkson, L. A., and E. Margaret Crawford. *Feast and Famine: Food and Nutrition in Ireland, 1500–1920*. New York: Oxford University Press, 2001.

Corrigan, D. J. *Famine and Fever as Cause and Effect in Ireland*. Dublin, 1846.

Davis, Mike. *Late Victorian Holocausts: El Niño Famines and the Making of the Third World*. New York: Verso, 2001.

Fenning, Hugh. "Typhus Epidemic in Ireland, 1817–19: Priests, Ministers, Doctors." *Collectanea Hibernica* 41 (1999): 117–39.

Flanagan, Thomas. *The Irish Novelists, 1800–1850*. New York: Columbia University Press, 1959.

Flinn, M. W. "The Poor Employment Act of 1817." *Economic History Review* 14 (1961): 82–92.

Gamble, John. *Views of Society and Manners in the North of Ireland, in a Series of Letters Written in the Year 1818*. London, 1819.

Harington, C. R., ed. *The Year without a Summer? World Climate in 1816*. Ottawa: Canadian Museum of Nature, 1992.

Harty, William. *An Historic Sketch of the Causes, Progress and Present State of the Contagious Fever Epidemic in Ireland*. Dublin, 1819.

Howard, Luke. *The Climate of London Deduced from Meteorological Observations, Made at Different Places, in the Neighbourhood of the Metropolis*. 2 vols. London, 1820.

Kidd, William. "A Concise Account of the Typhus Fever, at Present Prevalent in Ireland, as it Presented Itself to the Author in One of the Towns in the North of that Country." *Edinburgh Medical and Surgical Journal* 14 (1818): 144–58.

Kirkness, Ewen, et al. "Genome Sequences of the Human Body Louse and Its Primary Endosymbiont Provide Insights into the Permanent Parasitic Lifestyle." *Proceedings of the National Academy of Sciences* 107.27 (July 2010): 12168–73.

Malthus, Thomas. "Newenham and Others on the State of Ireland." *Edinburgh Review* (July 1808): 336–55.

O'Neill, Timothy P. "Fever and Public Health in Pre-Famine Ireland." *Journal of the Royal Society of Antiquaries of Ireland* 103 (1973): 1–34.

———. "The State, Poverty, and Distress in Ireland, 1815–45." Ph.D. thesis, University College, Dublin, 1984.

Parker, William. *A Plan for the General Improvement of the State of the Poor in Ireland*. Cork, 1816.

Peel, Robert. *Private Papers*. Ed. Charles Stuart Parker. London, 1891.

"Petition from Cork Complaining of the Increase of Poverty: From the Mayor, Sheriffs, & Several Inhabitants." *Hansard*, February 27, 1818.

Post, John D. *The Last Great Subsistence Crisis in the Western World*. Baltimore: Johns Hopkins University Press, 1977.

Raoult, Didier, et al. "The History of Epidemic Typhus." *Infectious Diseases Clinics of North America* 18 (2004): 127–40.

Raoult, Didier, and Véronique Roux. "The Body Louse as a Vector of Re-emerging Human Diseases." *Clinical Infectious Diseases* 29 (1999): 888–911.

Reid, Thomas. *Travels in Ireland*. London, 1823.

Report from the Select Committee on the State of Disease and Condition of the Labouring Poor in Ireland. London, 1819.

Robins, Joseph. *The Miasma: Epidemic and Panic in Nineteenth-Century Ireland*. Dublin: Institute of Public Administration, 1995.

Rogan, Francis. *Observations on the Condition of the Middle and Lower Classes in the North of Ireland, As It Tends to Promote the Diffusion of Contagious Fever, etc*. London, 1819.

Schneider, David P., et al. "Climate Response to Large, High-Latitude and Low-Latitude Volcanic Eruptions in the Community Climate System Model." *Journal of Geophysical Research* 114 (2009): D15101.

Sen, Amartya, and Jean Dreze, eds. *The Political Economy of Hunger*. Oxford: Clarendon Press, 1990.

Sharkey, Patrick. *An Essay on the Cause, Progress, and Treatment of Typhus Fever, Particularly as It Has Appeared in the City of Cork in the Years 1816 and 1817*. Cork, 1817.

Stern, W. M. "United Kingdom Public Expenditure by Votes of Supply, 1793–1817." *Economica* 17 (1950): 196–210.

Trant, Maurice. "Government Policy and Irish Distress, 1816–19." M.A. thesis, University College, Dublin, 1965.

Trotter, John. *Walks through Ireland in the Years 1812, 1814, and 1817.* London, 1819.

Webb, Patrick. "Emergency Relief during Europe's Famine of 1817: Anticipated Responses to Today's Humanitarian Disasters." *Journal of Nutrition* 132 (2002): 2092S–2095S.

CHAPTER NINE
HARD TIMES AT MONTICELLO

Anderson, Hattie. "Frontier Economic Problems in Missouri, 1815–28." *Missouri Historical Review* 34.1 (1939): 38–70.

Baldwin, Leland D. *The River Boat on Western Waters.* Pittsburgh: University of Pittsburgh Press, 1941.

Baron, William R. "1816 in Perspective: The View from the Northeastern United States." In *The Year without a Summer? World Climate in 1816,* ed. C. R. Harington, 124–44. Ottawa: Canadian Museum of Nature, 1992.

Bear, James A. Jr., ed. *Jefferson at Monticello.* Charlottesville: University of Virginia Press, 1967.

Bewell, Alan. *Romanticism and Colonial Disease.* Baltimore: Johns Hopkins University Press, 1999.

Birkbeck, Morris. *Notes on a Journey in America, from the Coast of Virginia to the Territory of Illinois, with Proposals for the Establishment of a Colony of English.* Ann Arbor: University of Michigan Press, 1968.

Brady, David M. "The Panic of 1819: Its Cause and Effects in Illinois History." Springfield: Sangamon County Historical Society, 2006.

Briffa, K. R., et al. "Tree-Ring Reconstructions of Summer Temperature Patterns across Western North America since 1600." *Journal of Climate* 5 (1992): 735–54.

Buffon, Comte de [Georges Louis Leclerc]. *Buffon's Natural History, Containing a Theory of the Earth, a General History of Man, of the Brute Creation, and of Vegetables, Minerals, etc.* 10 vols. London, 1797.

———. "Des époques de la nature." In *Histoire Naturelle,* 5:1–254. Paris, 1778.

Buley, R. Carlyle. *The Old Northwest: Pioneer Period, 1815–40.* Bloomington: Indiana University Press, 1950.

Cayton, Andrew R. L. "The Fragmentation of 'A Great Family': The Panic of 1819 and the Rise of the Middling Interest in Boston, 1818–22." *Journal of the Early Republic* 2 (Summer 1982): 143–67.

Chenoweth, Michael. "Daily Synoptic Weather Map Analysis of the New England Cold Wave and Snowstorms of 5 to 11 June, 1816." In *Historical Climate Variability and Impacts in North America,* ed. L. A. Dupigny-Giroux and C. J. Mock, 107–21. Berlin: Springer, 2009.

Dangerfield, George. *The Era of Good Feelings.* New York: Harcourt Brace, 1952.

Darby, William. *The Emigrant's Guide to the Western and Southwestern States and Territories*. New York, 1818.

Day, Clarence Albert. *A History of Maine Agriculture, 1604–1860*. Orono: Maine University Press, 1954.

Derks, Scott, and Tony Smith, eds. *The Value of a Dollar, 1600–1865*. Milberton, NY: Grey House Publishers, 2005.

Dorsey, Dorothy D. "The Panic of 1819 in Missouri." *Missouri Historical Review* 29.2 (1935): 79–91.

Dupre, Daniel S. "The Panic of 1819 and the Political Economy of Sectionalism." In *The Economy of Early America: Historical Perspectives and New Directions*, ed. Cathy Matson. University Park: Pennsylvania University Press, 2006.

———. *Transforming the Cotton Frontier: Madison County, Alabama, 1800–40*. Baton Rouge: Louisiana State University Press, 1997.

Fearon, Henry. *Sketches of America: A Narrative of a Journey of Five Thousand Miles through the Eastern and Western States of America*. London, 1818.

Fischer, David Hackett. *The Great Wave: Price Revolutions and the Rhythm of History*. New York: Oxford University Press, 1996.

Fleming, James R. *Historical Perspectives on Climate Change*. New York: Oxford University Press, 1998.

Flint, James. *Letters from America, Containing Observations on the Climate and Agriculture of the Western States, etc.* Edinburgh, 1822.

Gerbi, Antonello. *The Dispute of the New World: The History of a Polemic, 1750–1900*. Trans. Jeremy Mole. Pittsburgh: University of Pittsburgh Press, 1973.

Haulman, Clyde. *Virginia and the Panic of 1819: The First Great Depression and the Commonwealth*. London: Pickering & Chatto, 2008.

Hopkins, Edward J., and Joseph M. Moran. "Monitoring the Climate of the Old Northwest, 1820–95." In *Historical Climate Variability and Impacts in North America*, ed. L. A. Dupigny-Giroux and C. J. Mock, 171–88. Berlin: Springer, 2009.

Hoyt, Joseph B. "The Cold Summer of 1816." *Annals of the Association of American Geographers* 48 (1958): 118–31.

Hume, Edgar E. "The Foundation of American Meteorology by the United States Army Medical Department." *Bulletin of the History of Medicine* 8 (1940): 202–38.

Irving, Washington. *The Sketch Book of Geoffrey Crayon*. New York, 1819.

Jefferson, Thomas. *Farm Book*. Ed. Edwin Morris Betts. Princeton: Princeton University Press, 1953.

———. *Garden Book, 1766–1824*. Ed. Edwin Morris Betts. Philadelphia: American Philosophical Society, 1944.

———. *Notes on the State of Virginia*. Ed. William Peden. Chapel Hill: University of North Carolina Press, 1982.

———. *Papers*. Ed. Julian Boyd. 38 vols. Princeton: Princeton University Press, 1950–.

———. *Writings*. Ed. Paul Leicester Ford. 10 vols. New York: G. P. Putnam's Sons, 1892–99.

———. *Writings*. Ed. Andrew Lipscomb. 20 vols. Washington, DC: Thomas Jefferson Memorial Association, 1903–4.

Johnston, H.J.M. *British Emigration Policy, 1815–1830: "Shovelling Out Paupers."* Oxford: Clarendon Press, 1972.

Jones, P. D., et al. "Tree-Ring Evidence of the Widespread Effects of Explosive Volcanic Eruptions." *Geophysical Research Letters* 22.11 (June 1995): 1333–36.

Jones, Pomroy. *Annals and Recollections of Oneida County.* Rome, NY, 1851.

Kelvin, Lord [William Thompson]. "On the Secular Cooling of the Earth." *Transactions of the Royal Society of Edinburgh* 23 (1864): 167–69.

Ludlum, David. *Early American Winters, 1604–1820.* Boston: American Meteorological Society, 1966.

Malone, Dumas. *Jefferson and His Time.* 6 vols. Boston: Little, Brown, 1948–81.

Milham, Willis I. "The Year 1816—The Causes of Abnormalities." *Monthly Weather Review* 52.12 (December 1924): 563–70.

Post, John D. *The Last Great Subsistence Crisis in the Western World.* Baltimore: Johns Hopkins University Press, 1977.

Rezneck, Samuel. "The Depression of 1819–22: A Social History." *American Historical Review* 39.1 (1933): 28–47.

Rogers, Robert. *A Concise Account of North America.* London, 1765.

Rohrbough, Malcolm. *The Land Office Business: The Settlement and Administration of American Public Lands, 1789–1857.* New York: Oxford University Press, 1968.

Schoolcraft, Henry. *Personal Memoirs of a Residence of Thirty Years with the Indian Tribes on the American Frontiers, with Brief Notices of Passing Events, Facts, and Opinions, AD 1812 to AD 1842.* Philadelphia, 1851.

Schur, Leon. "The Second Bank of the United States and the Inflation after the War of 1812." *Journal of Political Economy* 68.2 (April 1960): 118–34.

Seybert, Adam. *Statistical Annals of the United States, 1789–1818.* Philadelphia, 1818.

Skeen, C. Edward. "'The Year without a Summer': An Historical View." *Journal of the Early Republic* 1 (Spring 1981): 51–67.

Stommel, Henry, and Elizabeth Stommel. *Volcano Weather: The Story of 1816, the Year without a Summer.* Newport, RI: Seven Seas Press, 1983.

Volney, Constantin-François, *View of the Climate and Soil of the United States of America.* London, 1804.

Webster, Daniel. *Private Correspondence.* Ed. Fletcher Webster. 2 vols. Boston: Little, Brown, 1857.

Williams, Samuel. *Natural and Civil History of Vermont.* 2nd ed. Burlington, 1809.

EPILOGUE
ET IN EXTREMIS EGO

Crutzen, Paul. "Albedo Enhancement by Stratospheric Injections: A Contribution to Resolve a Policy Dilemma?" *Climatic Change* 77 (2006): 211–19.

Fleming, James R. *Fixing the Sky: The Checkered History of Weather and Climate Control*. New York: Columbia University Press, 2010.

Lessing, Hans-Erhard. "What Led to the Invention of the Early Bicycle?" *Cycle History* 11 (2001): 28–36.

Moore, J. C., et al. "Efficacy of Geoengineering to Limit 21st Century Sea-Level Rise." *Proceedings of the National Academy of Science* 107.36 (September 2010): 15699–703.

Neely, R. R., et al. "Recent Anthropogenic Increases in SO_2 from Asia Have Minimal Impact on Stratospheric Aerosol," *Geophysical Research Letters* (published online, March 2013).

Robock, Alan, et al. "The Benefits, Risks, and Costs of Stratospheric Engineering." *Geophysical Research Letters* (2009): L19703.

Spielhagen, Robert F. "Enhanced Modern Heat Transfer to the Arctic by Warm Atlantic Water." *Science* 331 (January 2011): 450–53.

Vernier, J.-P., et al., "Major Influence of Tropical Volcanic Eruptions on the Stratospheric Aerosol Layer during the Last Decade." *Geophysical Research Letters* 38 (June 2011): L12807.

INDEX

Account of the Arctic Regions, An (Scoresby), *126*,
135, *136*

Adams, John Quincy, 224

Agassiz, Louis, 161, 167–68

agriculture, 29, 156; Alps and, 156, 164; ameliorative effects on climate of, 215; America
and, 203–6, 213–17, 221–28; barley, 61, 102, 113,
181; beans, 13, 102, 105, 115, 119; buckwheat,
113; Buffon and, 214; Byron and, 68; climate
change and, 216, 232; coffee, 13, 27, 81; corn,
13, 61, 176, 180, 203, 205–6; cotton, 13, 205,
221; droughts and, 9, 24, 39, 54, 57, 82–83, 87,
89–90, 102, 113, 139–40, 203, 205–6, 227–28;
Dustbowl and, 257n63; famine and, 62 (*see
also* famine); Ireland and, 179–80; Jefferson
and, 203, 205–6, 213–17, 221, 225–28; Little Ice
Age and, 39; Liverpool and, 62; Monticello
and, 203, 206, 217, 221, *223*, 224; oats, 61, 176,
178, 203, 206; pepper, 13, 27; potatoes, 61, 176,
180–81, 192, *194–95*; prices and, 179–80 (*see
also* prices); rice, 9–10, 13–14, 16–20, 23–24,
31, 71, 82, 90, 99, 101–5, 107–11, 113, 119; rye,
203, *222*; scientific progress in, 230; Russia
and, 10, 63, 198; subsistence, 9, 179; sugar, 27;
Sumbawa and, 13; Switzerland and, 62, 64;
wheat, 99, 102, 105, 110, 113, 115, 119, 164, 176,
203, 205–6, 214, 221, *222*, 226; Williams and,
215; Yunnan and, 99–103, 107–8, 117–20

AIDS, 93

algae, 89

Alps: Agassiz and, 161, 167–68; agriculture and,
156, 164; aquatic submersion and, 162; avalanches and, 46, 152; Chamonix and, 152–56;
Charpentier and, 161–62, 167–68, 251n24;
erratics and, 166; glaciers of, 129, 150–70;

global warming and, 156; Ice Age theory
and, 155–60, 162, 166, 168–69; ice dams and,
158, 159–60, 162–64, 166–67; *jökuljlaups* and,
159, 162, 170; Mount Le Pleureur and, 159,
161; Perraudin and, 160–62, 164, 166–68;
River Dranse and, 159–67; Shelleys and,
45–46, 151–59, 169–70, 206; Swiss Society
of Natural Sciences and, 157, 162, 166–67;
Tambora's effects on, 45–66, 150–51, 155–56,
159, 162–63, 165, 168–70; Val de Bagnes and,
159–68, 170; Venetz and, 157–58, 160–68, 170;
wheat and, 164; storms of, 45–46

America: agriculture and, 203–6, 213–17, 221–28;
Annsville snowstorm and, 200–201; Buffon's theories on, 206–16; climate change
and, 206–17, 227–28; climate optimism and,
126–27, 199–200, 212; climate pessimism
and, 206, 210–11, 226–28; Cumberland Pike
and, 225; drought and, 20–23, 205–6, 227–28;
Dustbowl of, 257n63; "Eighteen-Hundred-
and-Froze-to-Death" and, 9, 128, 200–206,
217; Era of Good Feelings and, 224; Erie
Canal and, 225; famine and, 205, 217,
224–25; Great Depression and, 225; Great
Lakes and, 203; Hamiltonian economics
and, 225; immigration to, 214, 218–19, 224; as
inhospitable continent, 210; Jefferson and,
200, 203, 205–7, 211–28; June 1816 weather
in, 203, 215–16; meteorology and, 199, 203,
211, 216–17, 227; Monticello and, 203, 206,
217, 221, *223*, 224; panic of 1819 and, 216–26;
poverty and, 218, 226; as Promised Land,
217; starvation and, 138, 205, 226; storms
and, 200–205, 227; Tambora's effect on, 71,
199–203, *204*, 216–27; War of Independence

America (*cont.*)
and, 1; wheat and, 203, 205–6, 214, 221, *222*, 226; Year without a Summer and, 199–200, 203, 225
American Indians, 212, 219
Andes, 34
animal mortality, 18, 178, 229
Annsville snowstorm, 200–201
Antarctica, 149
Anthropocene Period, 232–33, 258n4
Antoinette, Marie, 35
Arabian Sea, 78
Arabs, 26–27, 79, 86
Arctic: Atlantic Meridional Overturning Circulation (AMOC) and, 137–38, 140, 249n25; Baffin Bay and, 132, 141; Banks and, 121–22, 127–28, 135, 142; Barrow and, 122–34, 141–51; Bering Strait and, 128; British politics and, 121, 131, 148; Buchan and, *126*, 131, 133, 141, 144; Canada and, 40, 125, 133, 140–41, 144, 147; climate change and, 121–23, 130–31; Davis Strait and, 132; Denmark and, 128–29; *Frankenstein* and, 146–49, 250n31; Franklin and, 131–32, 144–49, 233; glaciers and, 129; global warming and, 122–31, 133, 146, 150, 258n5; Holocene Period and, 233; icebergs and, 127, 130, 151; Lancaster Sound and, 132, 141–42, 144; Little Ice Age and, 129, 131; Marine Diver and, 135–38; melting ice and, 121, 127, 129–33, 138, 140, 233, 258n5; Melville Bay and, 132; northeast passage and, 128; North Pole and, 48, 134, 137, 140, 147, 168; northwest passage and, 5, 123, 125, 127–29, 132–34, 142–49; Norway and, 129, 149; O'Reilly and, 131–34; Parry and, 131–32, 142–48; Ross and, *126*, 131–33, 141–42, 148, 233; Royal Navy and, 121–34, 141–51, 233; Royal Society and, 121, 128, 132; Scoresby and, 125–28, 131–38, 258n5; Tambora's effect on, 121–23, 132, 135, 138–41, 146, 149; von Kotzebue and, 128, 131, 141; whaling and, 125, 127, 135; wind and, 140
"Arctic Expeditions, The" (Porden), 144
Arctic Oscillation, 50–51
ash: Arabian Sea coast studies and, 78; blocked sun and, 3, 5, 22–23, 66–69; Crawford on, 15; fluorine and, 24, 48; P. B. Shelley and, 33; rain and, 12, 16–20, 22–25, 30, 48, *52*; sulfate aerosols and, 3, 37, *38*, 48, 78–79, 81, 121, 123, 141, 230; sulfur and, 5, 6, 43, 48, *52*, 79; Sumbawa and, 12, 15–25, 30; sunsets and, 2, 97; *zaman hujan au* (time of the ash rain) and, 24–25

Association for the Suppression of Mendicity, 191–92
Atlantic Meridional Overturning Circulation (AMOC), 137–38, 140, 249n25
Atlantic Ocean, 1–3, 10; Arctic Oscillation and, 50–51; Arctic studies and, 127, 130, 132, 135, 137–38, 140, 146; eastern U.S. and, 9, 42, 82, 98, 128, 138, 199–205, 210, 215–18, 220, 224, 226, 228; icebergs and, 127, 130, 132; Ireland and, 176; New England seaboard and, 9, 42, 82, 98, 128, 138, 199, 202–5, 215–16, 218, 224; transatlantic trade and, 64, 138, 210, 217, *222*, 224, 226, 228, 237n5; zones of, 15, 71, 79, 116, 226
atmosphere: circulation patterns in, 48–50, 82, 89, 130, 137–40, 249n25; National Center for Atmospheric Research (NCAR) and, 50–51, 81; precipitation patterns and, 49, 217; pressure variables and, 49–51, *80*, 99, 140, *177*, 202–3; tropopause and, 50; wind and, 49 (*see also* wind)
atmospheric pressure, 49–51, *80*, 99, 140, *177*, 202–3
Austen, Jane, 95, 142, 242n1
Australia, 11, 42, 134
automobiles, 229
avalanches, 46, 152, 218

bacteria, 88–90, 184
Baffin Bay, 132, 141
Bali, *14*, 19, 22–24
Banks, Joseph, 121–22, 127–28, 135, 142
Barker, Francis, 178, 191–92
barley, 61, 102, 113, 181
Barrow, John: Arctic and, 122–34, 141–51; bureaucratic achievements of, 133; *Frankenstein* and, 146–48; global warming and, 123–25, 127, 129–31, 133, 146; journalistic propaganda and, 124; Little Ice Age and, 131; meteorology and, 129–31; northwest passage and, 123, 125, 127–29, 132–34, 142, 144–46, 148–49; Parry and, 142–44; *Quarterly Review* and, 124–25, 129–30, 133, 142, 146–47; Ross and, 141; snubs O'Reilly, 133–34; snubs Ross, 141–42; snubs Scoresby, *126*
Bay of Bengal, 71, 78, 83, 89–91, 99
Bay of Naples, 35
beans, 13, 102, 105, 115, 119
beggars: Association for the Suppression of Mendicity and, 191–92; children and, 63, 172; poverty and, 9, *59*, 61, 63–65, 172, 188–93, 225

Benares (ship), 21

Bengal: Calcutta, 5, 73, *81*, 83–84, 86–87, 91, 96, 239n21; cholera and, 72–78, *81*, 83–96, 123, 178; famine and, 83; Ganges River and, 80–84, 88, 96; *ghauts* and, 81, 83–86; Hastings and, 74; monsoons and, 78–84, 88–91; Ola Bibi and, 83; rice and, 82, 90; *sharif* crop and, 81–82; storms and, 80–83, 99; Tambora's effects on, 71–72, 78–83, 90–91, 96

Bering Strait, 128

Betwah (River), 73, 76

Bible, 44, 49, 53, 57, 68, 121, 162, 165, 178

bicycles, 229

Birkbeck, Morris, 218

"Bitter Famine" (Li Yuyang), 113–14

Black Death, 94

black frost, 203

Black Prophet, The: A Tale of Irish Famine (Carleton), 173–74, *175*, 178, 180, *196*

Bologna Prophecy, 69–71

Borneo, 22

Bossons glacier, 153–55, 169

Brady, David M., 256n35

Brandes, Heinrich, 59

bread, 9, 61–62, *112*, 176

Brockedon, William, 164

Buchan, David, *126*, 131, 133, 141, 144

buckwheat, 113

Buffon, Comte de (Georges Louis Leclerc): American Indians and, 212; background of, 209; climate change and, 155, 206–16, 226–28; climate pessimism and, 226–28; earth science of, 130–31, 155, 206–16; Enlightenment and, 226; geo-engineering and, 226; "Great Climate Compromise" and, 215; *Histoire Naturelle* and, 208, 213; Jefferson and, 206–7, 211–16, 226–28; New World theories of, 206–16; temperature observations of, 209, 214

Burma, 91, 98, 119

Burns, Robert, 172

Burundi, 183

Byron, George Gordon, 9, 116; apocalyptic poetry of, 68–69; Arctic and, 148; *Childe Harold's Pilgrimage* and, 46–47, 54, 125; Clairmont and, 45, 51; "Darkness" and, vii, 67–69; death by malaria of, 76–77; famine and, 69; ghost story contest and, 51, 57, 156; global warming and, 156; Polidori and, 51; storms and, 68, 77; Switzerland and, 45–47, 51, 65–67, 69; Tambora and, 47; vampires and, 52–53; Villa Diodati and, 46, 51, 67, 68

calderas, 5–6, 7, 11, 21

Cambodia, 14

Canada: Arctic and, 40, 125, 133, 140–41, 144, 147; cold fronts of, 217; Hudson Bay and, 140–41, 147; immigrants and, 219; snowstorm of 1816 and, 202; Tambora's effects on, 202, 232

Carleton, William: *The Black Prophet* and, 173–74, *175*, 178, 180, *196*; despair over countrymen, 173–75, 178; Irish famine and, 173–74, *175*, 178, 180, 184, 186, 188, 190, 192, 194–95, *196*; typhus and, 184, 186, 188, 190, 192

Casa Galetti, 76

Celebes. *See Sulawesi*

Charpentier, Jean de, 161–62, 167–68, 251n24

Childe Harold's Pilgrimage (Byron), 46–47, 54, 125

children: Annsville storm and, 200–201; beggar, 63, 172; disease and, 74, 188; famine and, 5, 9, 23, 25, 63–65, 98, 109–14, 198

China, 14, 138, 199; "century of humiliation" of, 106; cholera and, 91; Confucianism and, 98, 106–7, 109–10, 114, 120; Han Chinese and, 100–101, 103; Himalayan plateau and, 82, 99, 102, 115; northwest passage and, 125; opium and, 115–20, 123; Qing dynasty and, *100*, 101, 105–13, 117, 120; Yunnan and, 9, 11, 97–120

cholera, 5; actions of, 77; bacteria and, 88–90; Bengal and, 72–78, *81*, 83–96, 123, 178; Calcutta Medical Board and, 87; climate change and, 86–91, 94; Colwell and, 89–90; deities of, 74–5, 83; etiology for, 183; famine and, 9; fecal matter and, 88; Ganges River and, 80–84, 88, 96; global epidemic of, 71, 91–95, 123, 230, 233, 236; Grand Army and, 72–76; great fear of, 75–76, 85–86, 92, 94; International Sanitary Conference and, 93; Jameson and, 87–88, 91, 95, 178; Koch and, 87–88; Medwin on, 76–77; meteorology and, 87–91; nineteenth-century literature on, 72; Pascual and, 89–90; pathogens of, 87–90; plankton and, 88–90; recent research on, 246n25; superstition and, 74–75; temperature and, 90

Christmas Carol, A (Dickens), 40

circulation patterns: atmospheric, 48–50, 82, 89, 130, 137–40; meridional, 48–50, 82, 137–38, 249n25; ocean, 130, 140, 146; thermohaline, 137, 140

Clairmont, Claire, 45, 51, 76, 115

climate: ash and, 67 (*see also* ash); Atlantic Meridional Overturning Circulation

climate (*cont.*)
(AMOC) and, 137–38, 140, 249n25; atmospheric pressure and, 49–51, *80*, 99, 119, 136, 140, *177*, 202–3; Calcutta Medical Board and, 87, 91; circulation patterns and, 48–50, 82, 89, 130, 137–40, 146, 249n25; coldness of 1810s and, 36–42; droughts and, 9, 24, 39, 54, 57, 82–83, 87, 89–90, 102, 113, 139–40, 203, 205–6, 227–28; Earth's tilt and, 49, 80; El Niño and, 82, 90; eruption of 1258 and, 41, 43–44, 216; eruption of 1809 and, 37–39, 41–43, 58, 216, 241n12; floods and, 9, 20, 39, 46, 54, 56–57, 59–60, 83, 89–90, 102, 105, 107–8, 121, 159–60, 162, 164–67, 169–70, 218, 227–28; "Frankenstein weather" and, 91, 107, 174, 227–28; Franklin's observation of altered, 1–2; frost and, 9, 37, 39, 58, 60, 98, 102, 105, 127, 130, 142, 147–48, 155, 200–206, 210, 233; glaciers and, 37, 55, 129, 150–70; heat transfer and, 49, 50, 137–38, 140, 249n25; Himalayan, 82, 99, 102, 115; Holocene Period and, 42, 44, 232–33, 241n12; ice core analysis and, 37, 43, 48, 241n12; jet streams and, 50, 203; latitude and, 2, *3*, 43, 49–50, 56, 79–80, 91, 98, 101, 122, 130, 137, 139, 150, 203, 209, 234; moderate, 42, 49, 100, 102, 137, 212, 215, 226, 228, 230; optimism and, 126–27, 199–200, 212; pessimism and, 206, 210–11, 226–28; precipitation patterns and, 49, 217; snow and, 1, 37, 42, 45–46, 56–58, 60–61, 67, 98, 102, 105, 115, 130, 144, 147, 151–52, 154, 157, 163, 200–203, 205, 210, 215; solar radiation irregularity and, 40–41; stratospheric effects and, *3*, 8, *19*, 43, 48, 50, 52, 67, 79, 90, 115, 230; sulfate aerosols and, *3*, 37, *38*, 48, 78–79, 81, 121, 123, 141, 230; sunspots and, 40–41, 70; Tambora's effects on, 8 (*see also* Tambora); trade winds and, *19*, *23*, 78, 82, 99; tree-ring studies and, 24, 43, 82–83, 217, 248n19; volcanism and, 6, 10, 26, 33–34, 43, 235, 238n10 (*see also* volcanoes); weather patterns and, 37, 43, 55, 59, 67, 102, 226; wind and, 49–51, 150, 173, 178; Yunnan and, 98–105; zones of, 10, 15, 71, 79, 82, 91, 99, 138, 216
climate change: agriculture and, 216, 232; America and, 206–17, 227–28; Arctic and, 121–31, 133, 146, 150, 258n5; Buffon and, 226–28; cholera and, 86–91, 94; Crutzen and, 230, 232; fossil fuels and, 123, 226, 232; geo–engineering solutions to, 230–31; glaciers and, 152, 158, 165–70; global warming and, 123–31, 133, 146, 150, 156, 228, 231, 258n5;

"Great Climate Compromise" and, 215; human affairs and, 5–6, 8, 10; Intergovernmental Panel on Climate Change (IPCC) and, 235; Ireland and, 180–81, 184, 195; Little Ice Age and, 41; Medieval Warming Period and, 40; melting ice and, 121, 127, 129–31, 138, 140, 233, 258n5; modern concerns of, 230–35; Pinatubo and, 138–39; refugees and, 216, 218; Royal Society and, 121; scale of threat of, 233; social consequences of, 24, 36, 41, 61–62, 64, 67–69, 71, 86–87, 95, 98, 110, 112, 117, 120, 171, 180, *182*, 194–95, 197, 199–200; Sumbawa and, 24; Swiss Society for Natural Science and, 157; trauma from, 59–60, 67, 71, 230–31; Venetz and, 157–58, 160–68, 170; volcanism and, 6, 10, 26, 33–34, 43, 235, 238n10; Yunnan and, 119–20
climate models, 40, 81
Climate of London, The (Howard), 55, 58–59
coffee, 13, 27, 81
Cold War, 3
Coleridge, Samuel Taylor, 39, 51, 60, 70, 99
colonialism: America and, 205, 209–10, 212; Bengal and, 85, 87, 93–94; Ireland and, 172, 191; Sumbawa and, 17, 27, 29–30; typhus and, 183
Colwell, Rita R., 89–90
Committee on Manufactures, 225
Confucianism, 98, 106–7, 109–10, 114, 120
Constable, John, 40, 50–51, 60, 174
Cook, James, 128
Corbyn, Frederick, 74
Corinne (de Staël), 34
corn, 13, 61, 176, 180, 203, 205–6
Cortez, Hernán, 128
Cotopaxi, 34
cotton, 13, 205, 221, 233
County Tyrone, 5, 173, 181, 186, 190, 193
Crater Lake, 21
Crawfurd, John, 15
Croker's Mountains, 142, 144
Crutzen, Paul, 230, 232
Cuvier, Georges, 130
cyclones, 50, *80*

"Darkness" (Byron), vii, 67–69
Darwin, Charles, 135, 232
Davis Strait, 132
Davy, Humphry, 34
Day of Judgment, 70
deforestation, 24–25, 214, 234

Denmark, 128–29
de Staël, Madame, 34
Dickens, Charles, 40
disease: bacterial, 88–90, 184; Black Death, 94; children and, 188; cholera, 9, 72–78, *81*, 83–96, 123, 178, 183, 230, 233, 236, 246n24, 246n25; epidemics, 71–75, 81, 84–96, 123, 181–97, 230, 233, 236; fecal matter and, 88; Great Famine and, 173, 194, 198 (*see also* famine); International Sanitary Conference and, 93; lice and, 176, 179, 181–83; malaria, 76–77; pathogens and, 87–90, 180, 183–84; plagues, 44, 78, 92, 95, 183, 190–91; riots and, 94; syphilis, 183; tuberculosis, 77, 221; typhus, 9, 66, 69–70, 176, 179, 181–93, 198; yellow fever, 183
Dranse (River), 159–67
Drayton, Michael, 212
drinking water, 7
droughts, 9; America and, 203, 205–6, 227–28; Bengal and, 82–83, 87, 89–90; Howard and, 57; Java and, 24; Little Ice Age and, 39; Masters and, 54; Pinatubo and, 139–40; Scandinavia and, 57; Yunnan and, 102, 113
dry season, 13, 79, *80*, 86
Dublin Evening Post newspaper, 190
Dublin Royal Society, 132
Duke of Wellington, 12
Dupre, Daniel, 223
Dustbowl, 257n63
Dutch: Bengal and, 79; cholera and, 86; Sumbawa and, 13–15, 20, 24–25, 27, 29–30, 62

earthquakes, 22; Byron and, 47; New Madrid, 34; Shelley and, 33
earth science: Buffon and, 130–31, 155, 206–16; Cuvier and, 130; Humboldt and, 33
East India Company, 21, 27, 29
Ecuador, 231
Egypt, 44
"Eighteen-Hundred-and-Froze-to–Death," 9, 128, 200–206, 217
Elizabeth, Queen of England, 128
El Niño, 82, 90
England: Act of Union and, 188 (*see also* Ireland); Arctic region and, 127, 129–30, 141; cholera and, 92; famine riots and, 61; Grand Army and, 72–76; Howard's climate observations and, 55–57, 59; Little Ice Age and, 34, 37, 40; opium and, 115–20; Pentrich Revolution and, 61–62; Royal Navy and, 20, 121–34, 141–51; storms and, 60; Tambora's

effect on, 57–60; typhus and, 253n43; Waterloo and, 70, 116, 148, 180; welfare politics in, 197; Weymouth Bay, 50
Era of Good Feelings, 224
Essay on the Modification of Clouds (Howard), 55
evacuations, 5, 6, 8, 163
Evening Post newspaper, 190

famine, 230; America and, 205, 217, 224–25; Bengal and, 83; *The Black Prophet* and, 173–74, *175*, 178, 180, *196*; "Bread or Blood" slogan and, 61; Byron on, 69; Carleton's writings and, 173–74, *175*, 178, 180, 184, 186, 188, 190, 192, 194–95, *196*; children and, 5, 9, 23, 25, 63–65, 98, 109–14, 198; cholera and, 9; Europe's last, 60–64; Great Famine (Ireland), 173, 194, 198; Ireland and, 171–74, 176, 179–81, *182*, 184–90, 193–98; Lord Liverpool's response to, 62; media and, 190; opium and, 115–20; Pentrich Revolution and, 61–62; poetry on, 106–15; Raffles and, 30–31; refugees and, 63–64, 98, 185, 187–89; riots and, 9–10, 61–62, 193–94; Russell and, 174; starvation and, 5, 9, 19, 23–24, 61–64, 69–70, 105, *112*, 113–14, 138, 144–45, 148, 155, 172–73, 181, 187, 189, 194, 198, 205, 226, 233–34, 239n14; Sumbawa and, 24, 30; supply and demand effects on, 112; Switzerland and, 62–66, 69; Trevelyan and, 174; typhus and, 9, 70; Yunnan and, 102–20
fecal matter, 88
floods, 9; Alps and, 159–67, 169–70; America and, 227–28; Bengal and, 83; cholera and, 89–90; dykes and, 59; Howard and, 56–57, 60; *jökulhlaups* and, 159, 162, 170; Little Ice Age and, 39; Masters and, 54; River Dranse and, 159–67; Sumbawa and, 20; Switzerland and, 46; Val de Bagnes and, 159–68, 170; "Year of the Beggar" and, 59; Yunnan and, 102, 105, 107–8, 121
Flores island, 14, 18, *23*
fluorine, 24, 48
Forster, Thomas, 2–3, *6*–7, 55
fossil fuels, 123, 226, 232
France, 9, 12, 58–59, 62–63, 211
Frankenstein (Mary Shelley): Arctic and, 146–49, 250n31; Barrow and, 146–48; as cautionary tale, 234; cultural significance of, 146; refugees and, 64–66; Shelley's inspiration for, 9, 45, 53–54, 59, 65–67, 95–96, 125, 142, 146–49, 156–57, 170, 195, 197, 210, 234
"Frankenstein weather," 91, 107, 174, 227–28

Franklin, Benjamin, 1–2, 5, 7
Franklin, John: Arctic explorations and, 131–32, 144–49, 233; books on, 259n28; death of, 145–46; as "The Man Who Ate His Boots," 144
free trade, 30
French Revolution, 35–36
Friedrich, Caspar David, 3
frost: 9, 37, 39, 58, 60, 98, 102, 105, 127, 130, 142, 147–48, 155, 200–206, 210, 233; black, 203; early, 60; late, 37
Fuji, Mount, 8

Gallatin, Albert, 205
Gamble, John, 186–87
Ganges (River), 80–84, 88, 96
Gaullieur, Eusèbe–Henri, 64
George III, King of England, 1
Germany, 3, 59, 61, 197
ghauts (stone steps), 81, 83–86
Giétro glacier, *158*, 159–61, 163–65, 170
glaciers: advancement of, 151–55, 157, 159, 162, 169; Agassiz and, 161, 167–68; Alps and, 129, 150–70; Arctic and, 129; Bossons, 153–55, 169; Charpentier and, 161–62, 167–68, 251n24; climate and, 37, 55, 129, 150–70; Giétro and, *158*, 159–61, 163–65, 170; Ice Age theory and, 155–60, 162, 166, 168–69; ice dams and, *158*, 159–60, 162–64, 166–67; *jökulhlaups* and, 159, 162, 170; Mer de Glace and, 154, 156; Mont Blanc, 152–57, 159, 169–71; Perraudin and, 160–62, 164, 166–68; starvation and, 155; Swiss Society of Natural Sciences and, 157, 162, 166–67; temperature and, 152, 157, 159–60; Venetz and, 157–58, 160–68, 170
Global Volcanism Program, 42
global warming. *See* climate change
Godwin, Fanny, 45–46, 115–16
Godwin, Mary. *See* Shelley, Mary
Godwin, William, 96
Goethe, Johann Wolfgang von, 55
Grand Army (India), 72–76
Grand Banks, 127
Grand Tour, 34
Grant, Charles, 192–94, 197
"Great Climate Compromise," 215
Great Depression, 225
Great Famine (Ireland), 173, 194, 198
Great Lakes, 203
Greenland, 40, 202, 232; Arctic exploration and, 125, 127, 129, 132–34, 137–38, 140–42; Baffin Bay, 132, 141; Davis Strait, 132; Godthaab

(Nuuk), 138; ice core studies and, 37; Lancaster Sound, 132, 141–42, 144; Spitsbergen and, 138; sulfate deposits in, *38*; whaling and, 125
Greenland (O'Reilly), 133–34

hail, 58–59, 83, 163, 179, 203, 236
Hamilton, William, 34
Hamlet (Shakespeare), 123
Hamlin, Christopher, 93
Hampstead Heath, 60
Han Chinese, 100–101, 103
Hardaul Lala, 74–75
Harty, William, 186, 190–91
Hastings, Lord, 72–76, 77
Heine, Heinrich, 92
Himalayas, 82, 99, 102, 115
Hinduism, 28, 82
Histoire Naturelle (Buffon), 208, 213
History of Java (Raffles), 29, 31, 32
Holocaust, 24
Holocene Period, 42, 44, 232–33, 241n12
homelessness, 66, 171, *182*, 189, 234
Homer, 72
horses, 14, 18, 20–24, 58, 61, 178, 211, 229
Howard, Luke, 55–60, 174, 243n17
Huaynaputina, 44
Hudson Bay, 140–41, 147
Humboldt, Alexander von, 33
Hunt, Leigh, 124–25
hurricanes, 32, 58
Hutton, James, 168

Ice Age theory, 230; Alps and, 155–60, 162, 166, 168–69; Little Ice Age and, 33–44, 129, 131
icebergs, 127, 130, 151
ice core analysis, 37, 43–44, 48, 241n12
Ice-Core Volcano Index (IVI), 43–44
ice dams, *158*, 159–60, 162–64, 166–67
ice floes: Alps and, 150–70; Arctic and, 130
Iceland, 1–2, 7, 34, 40, 43, 50–51, 159
Iliad (Homer), 72
India: Calcutta, 5, 73, *81*, 83–84, 86–87, 91, 96, 239n21; cholera and, 72–78, *81*, 83–96, 123, 178; Ganges River and, 80–84, 88, 96; *ghauts* and, 81, 83–86; Hastings and, 72–76, 77; Hinduism and, 28, 82; monsoons and, 78–84, 88–91; Ola Bibi and, 83; opium and, 115–20; *sharif* crop and, 81–82; Tambora's effects on, 71–72, 78–83, 90–91, 96
Indian Ocean, 22, 78, 80, 82, 89, 98, 102
Indonesia, 6, 8–9, 92, 199, 231

insects, 89, 170, 183
Intergovernmental Panel on Climate Change (IPCC), 235
International Sanitary Conference (1851), 93
Ireland: Act of Union and, 188; agriculture and, 179–80; Association for the Suppression of Mendicity and, 191–92; Board of Health and, 197; Carleton and, 173–74, 175, 178, 180, 184, 186, 188, 190, 192, 194–95, 196; climate change and, 180–81, 184, 195; conditions of 1817 in, 179–81, 185–94, 197–98; Cork, 180, 185–86, 189, 194; County Tyrone, 5, 173, 181, 186, 190, 193; Dublin, 132, 150–51, 173, 176, 178–79, 189–94; famine and, 171–74, 176, 179–81, 182, 184–90, 193–98; Gamble and, 186–87; homelessness and, 171, 182, 189; Irish character and, 187–98; Keats and, 171–73; landlords and, 173, 185; lice and, 176, 179, 181–87; Limerick, 188, 194; meteorology in, 150–51, 180, 190, 192, 196; Poor Employment Act and, 197; potatoes and, 176, 180–81, 192, 194–95; poverty and, 172, 176, 179–80, 186–87, 191; riots and, 193–94; storms and, 176, 177, 179; Tambora's effect on, 71, 171–74, 178–81, 186, 189–90, 193–95, 197–98; typhus and, 66, 176, 179, 181–93, 198; wheat and, 176; winds of, 178; "Year without a Summer" and, 174, 176, 177
Israelites, 44
Israel (official), 17

Jackson, Andrew, 225
Jacobins, 35, 191
Jameson, James, 87–88, 91, 95, 178
Jameson, Robert, 135
Japan, 8, 91
Java, 8, 134, 239n15; cholera and, 91; Raffles and, 27–32, 62, 240n33; Sumbawa and, 13–14, 16, 19, 22–24, 27–32
Java Trench, 8
Jefferson, Thomas: agriculture and, 205–6, 213–17, 221, 225–28; American Indians and, 212; Buffon and, 206–7, 211–16, 226–28; climate pessimism and, 226–28; drought and, 205; "Eighteen-Hundred-and-Froze-to-Death" and, 205–6, 217; Gallatin and, 205; "Great Climate Compromise" and, 215; Hamiltonian economics and, 225; meteorological studies of, 201, 211, 216–17, 227; Monticello and, 203, 206, 217, 221, 223, 224; Notes on the State of Virginia and, 211–13, 215; panic of 1819 and, 216–26

jet streams, 50, 203
Jews, 24
Jiaqing, Emperor, 109–10
jökulhlaups (glacier-floods), 159, 162, 170
jugement dernier des rois, Le (Day of Judgement for the Kings) (Maréchal), 36
Jura mountains, 46

Katmai, Mount, 42
Keats, George, 221
Keats, John, 39, 60, 76–77, 171–73, 221
Keats, Tom, 172
Kilkenny Moderator newspaper, 190
Klingaman, Nicholas P., 237n5
Klingaman, William K., 237n5
Koch, Robert, 87–88
Krakatau, 8, 37, 41–42, 44, 53
"Kubla Khan" (Coleridge), 99
Kun Hua Five, 107
Kuwae, 37, 44

Lake Geneva, 46, 53, 55, 67, 164
Laki (Lakagígar), 2, 43
Lamb, H. H. (Hubert), 50, 177
Lancaster Sound, 132, 141–42, 144
landlords, 173, 185
Laos, 98, 119
Last Man, The (Mary Shelley), 95–96
latitude: Atlantic Meridional Overturning Circulation (AMOC) and, 137–38, 140, 249n25; climate and, 2, 3, 43, 49–50, 56, 79–80, 91, 98, 101, 122, 130, 137, 139, 150, 203, 209, 234; Earth's tilt and, 80; high, 3, 43, 56, 122, 139; mid, 3, 49, 203; northern, 122; southern, 130, 150; temperature and, 137, 209; tropical, 2, 79, 98 (see also tropics); Volcanic Explosivity Index (VEI) and, 43
leeches, 20, 26
Le Pleureur, Mount, 159, 161
Letters from an American Farmer (St. John Crèvecoeur), 212
lice: host–parasite relationship and, 181, 183; propagation of, 181, 183; typhus and, 176, 179, 181–87
lightning, 53–54, 57–59, 115, 179
Little Ice Age: agriculture and, 39; Arctic and, 129, 131; Barrow and, 131; climate change and, 41; coldness of 1810s and, 36–42; Ice–Core Volcano Index (IVI) and, 43–44; intermittent nature of, 40; northwest passage and, 129; poetry and, 34, 39; Robock and, 40–41; tropics and, 37, 39; volcanic causes of, 40–41

Li Yuyang: "Bitter Famine" and, 113–14; famine poetry of, 11, 106–15, 120; official biography of, 247n11; "A Sigh for Summer Rain" and, 107–8; tax collectors and, 108–10; translations of, 236, 248n12
Lombok, 8, *14*, 22–23, 25, 236

Mackenzie, George Steuart, 34
Mackenzie, George, 39
magma, 6, 17–21, 43
Maison Chappuis, 45–46, 115–16
malaria, 76–77
Manchester Philosophical Society, 2
Mansfield, Calvin, 205
Maratha War (1817–18), 73, 74, 77
Maréchal, Sylvain, 36
Marine Diver, 135–38
Masters, Jeff, 54–55
McClure, Robert, 146
Medieval Warming Period, 40
Medwin, Thomas, 76–77, 245n7
Mekong delta, 98, 102, 119
melting ice, 121, 127, 129–33, 138, 140, 233, 258n5
Melville, Lord, 128
Melville Bay, 132
merchants, 27, 79, 86, 94, 96, 189, 194
Mer de Glace glacier, 154, 156
meridional circulation patterns, 48–50, 82, 137–38, 249n25
"Meteorological Imaginations and Conjectures" (Franklin), 1–2
Meteorological Register, 227
meteorology, 67, 80, 96, 230; America and, 199, 203, 211, 216–17, 227; Barker and, 190, 192; Barrow and, 129–31; Brandes and, 59; Carleton and, *196*; cholera and, 87; Colwell and, 89–90; El Niño and, 82, 90; Forster and, 2–3, 6–7, *55*; Franklin and, 1–2; Grant and, 192; Howard and, 55–60, 174; Ireland and, 150–51, 180, 190, 192; Jameson and, 87–88, 91, 95, 178; Jefferson and, 203, 211, 216–17, 227; Krakatau and, *53*; Laki and, 2; Mackenzie and, 39; Masters and, 54–55; Milham and, 199; NASA and, 40; Pascual and, 89–90; Yunnan and, 97–99, 104–5, 109
Milham, Willis, 199
Minoans, 44
monsoons: China and, 99, 102; India and, 78–84, 88–91; *sharif* crop and, 81–82; Sumbawa and, 22; Yunnan and, 99, 102
Mont Blanc, 152–57, 159, 169–71
"Mont Blanc" (P. B. Shelley), 153–55, 157, 169–71

Monticello, 203, 206, 217, 221, *223*, 224
Montserrat, 231
Morning Chronicle newspaper, 150–1
mortality rates, 63, 233, 254n45

Napoleon Bonaparte, 64, 124–25, 151; Alps and, 152; Napoleonic Wars and, 5, 7, 9, 56, 179; typhus and, 183; Waterloo and, 70, 116, 148, 180
NASA, 40
National Center for Atmospheric Research (NCAR), 50–51, 81
Natural and Civil History of Vermont (Williams), 215–16
natural selection, 232
Nelson, Horatio, 128–29, 144
New England, 9, 42, 82, 98, 128, 138, 199, 202–5, 215–16, 218, 224
New Hampshire Patriot newspaper, 205
New Madrid earthquakes, 34
northeasters, 46
northeast passage, 128
North Pole, 48, 134, 137, 140, 147, 168
northwest passage, 5, 123, 125, 127–29, 132–34, 142–49
Norway, 129, 149
Notes on the State of Virginia (Jefferson), 211–13, 215

oats, 61, 176, 178, 203, 206
ocean circulation, 130, 140, 146
Ola Bibi, 83
opium, 51, 99, 115–20, 123, 230
Oppenheimer, Clive, 238n10, 239n14
O'Reilly, Bernard, 131–34

Pacific Ocean, 8, 125, 132, 141–42, 146
paleoclimatology: Canadian Arctic and, 40; ice core analysis and, 37, 43, 48, 241n12; Iceland and, 40; Volcanic Explosivity Index (VEI) and, 43
Panic of 1819, 216–26
Papua New Guinea, 231
Parker, William, 185
Parry, William Edward, 131–32, 142–48
Pascual, Mercedes, 89–90
pathogens, 87–90, 180, 183–84
Peacock, Thomas Love, 33, 153
Peel, Robert, 176, 197
Pentrich Revolution, 61–62
pepper, 13, 27
Perraudin, Jean-Pierre, 160–62, 164, 166–68
Pfister, Christian, 155

Philippines, 42, 48, 52, 91, 138–39, 241n22
Phillips, Owen, 20, 31
Pierre à Bot, 166
Pinatubo, Mount, 42, 48, 52, 138–39, 241n22
pirates, 15–16, 21
Place de la Révolution, 35
plankton, 88–90
plinian eruptions, 18–20, 22, 23
Pliny the Younger, 18
poetry: Byron and, 9 (see also Byron, George
    Gordon); Coleridge and, 39, 51, 60, 70, 99;
    Confucian, 120; famine and, 11, 106–15, 120,
    236; Goethe and, 55; Homer and, 72; Keats
    and, 39, 60, 76–77, 171–73, 221; Little Ice Age
    and, 34, 39; Li Yuyang and, 11, 106–15, 120,
    236; Porden and, 144; Qing-era, 105–13, 120;
    Romantic Age and, 34; of Seven Sorrows,
    11, 120, 234; Shakespeare and, 72; Shelley
    and, 9 (see also Shelley, Percy Bysshe); Sum-
    bawa and, 17, 37; Tennyson and, 146; Wang
    Can and, 11; Yunnan and, 106–15
polar bears, 135, 233
Polidori, John William, 51–52
Polo, Marco, 99
Pompeii, 18, 33, 37, 44
Poor Employment Act, 197
Porden, Eleanor, 144
Portuguese, 27, 79
Post, John, 62
potatoes, 61, 176, 180–81, 192, 194–95
Poussin, Nicolas, 232–33
poverty: America and, 218, 226; Association for
    the Suppression of Mendicity and, 191–92;
    beggars and, 9, 59, 61, 63–65, 172, 188–93, 225;
    Carleton's writings and, 173–74, 175, 178, 180,
    184, 186, 188, 190, 192, 194–95, 196; homeless-
    ness and, 66, 171, 182, 189, 234; Ireland and,
    172, 176, 179–80, 186–87, 191; Irish landlords
    and, 173, 185; panic of 1819 and, 216–26; Poor
    Employment Act and, 197; Sumbawa and,
    25; Yunnan and, 107
precipitation patterns, 49, 217
prices, 27; bread, 9, 61–62; carbon waste and,
    226; commodity, 179, 227; cotton, 221; flour,
    220–21; grain, 10, 63, 110, 155, 179–80, 197–98,
    221, 222, 224; land, 219; rice, 9, 105, 111; slav-
    ery and, 23; transatlantic trade and, 64, 138,
    210, 217, 222, 224, 226, 228, 237n5
Pride and Prejudice (Austen), 95
Prussia, 197
pumice, 17, 21–22, 33
pyroclastic flows, 18–20, 22, 170

Qing dynasty, 100, 101, 105–13, 117, 120
Quarterly Review, 124–25, 129–30, 133, 142, 146–47
Queen Mab (P. B. Shelley), 130, 210

Raffles, Stamford, 37, 63, 239n11; as colonial
    anthropologist, 29; East India Company
    and, 29–30; famine aid and, 30–31; Java
    and, 27–32, 62, 240n33; as "philosopher
    king," 27–32; Singapore and, 24; slavery and,
    23–24, 239n21
Raffles, Thomas, 62
railroads, 229
raja of Sanggar, 13–20
refugees: climate change and, 216, 218; famine
    and, 63–64, 98, 185, 187–89; Frankenstein and,
    64–66
Reign of Terror, 35
Rhône (River), 164
rice, 9–10; Bengal and, 82, 90; selling humans for,
    23; Yunnan and, 71, 99, 101–5, 107–11, 113, 119
Ring of Fire, 8
Riot Act, 61
riots: disease and, 94; famine and, 9–10, 61–62,
    193–94; in Ireland, 193; Lord Liverpool's
    response to, 62; Pentrich Revolution and,
    61–62
Robertson, William, 210
Robespierre, 35
Robock, Alan, 40–41
Romanticism, 33 34, 35, 76–77, 95, 156, 169
Ross, John, 126, 131–33, 141–42, 148, 233
Royal Navy, 20; Arctic and, 121–34, 141–51;
    Barrow and, 121–34, 141–51; Buchan and,
    126, 131, 133, 141, 144; Franklin and, 131–32,
    144–49, 233; Lord Melville and, 128; Nelson
    and, 128–29, 144; northwest passage and,
    123, 125, 127–29, 132–34, 142, 144–46, 148–49;
    O'Reilly and, 134; Parry and, 131–32, 142–48;
    Ross and, 126, 131–33, 141–42, 148, 233
Royal Society, 121, 128, 132
Russell, Lord John, 173–74
Russell, Ken, 51
Russia, 48; agriculture and, 10, 63, 198; Arctic
    and, 128, 232; Bering Strait and, 128; chol-
    era and, 86, 92; grain and, 63, 198; immi-
    grants and, 64; northeast passage and, 128;
    typhus and, 183; von Kotzebue and, 128, 131,
    141, 183
rye, 203, 222

Saint Helens, Mount, 8, 43
St. John Crèvecoeur, 212

St. John's Chapel, 51

Sanggar Peninsula, 7, 13–24, 170; raja of, 13–20

Santorini, 44

Schoolcraft, Henry, 219

Scoresby, William: *An Account of the Arctic Regions* and, *126*, 135, *136*; Arctic and, 125, *126*, 127–28, 131–38, 258n5; Marine Diver of, 135–38; polar ice treatise of, 125; snubbed by Barrow, *126*; whaling and, 125, 127, 135

Scotland, 37, 56, 62, 88, 172

Scott, Robert, 149

Sen, Amartya, 190

Shakespeare, William, 72, 123

sharif crop, 81

Shelley, Mary, 47; Alps tour and, 45–46, 151–59, 169–70, 206; Arctic and, 124–25, 131, 134, 146–48; Barrow and, 124–25, 131, 146–47; Bath and, 116; biographers' depictions of, 65; Byron and, 46, 52, 54, 66, 69; de Staël and, 34; *Frankenstein* and, 9, 45, 52–55, 59, 65–67, 95–96, 125, 142, 146–49, 156–57, 170, 195, 197, 210, 234; ghost story contest and, 51, 57, 156; Hunt and, 124–25; *The Last Man* and, 95–96; literary friends of, 9, 51, 55; Pompeii and, 33–34, 37; reading Buffon, 210; sisters of, 115; Switzerland and, 45–46, 51–52, 54–55, 57, 62, 151; Walton character and, 147

Shelley, Percy Bysshe, 9, 47; Alps tour and, 45–46, 151–59, 169–70, 206; beggar children and, 65; Byron and, 9; death of, 77, 95; de Staël and, 34; Fanny Godwin and, 116; ghost story contest and, 51, 57, 156; global warming and, 130; Medwin and, 76; "Mont Blanc" and, 153–55, 157, 169–71; near drowning of, 53; Peacock and, 33, 153; poetry of as "satanic," 139; Pompeii and, 33; *Queen Mab* and, 130, 210; Switzerland and, 45, 51–52, 65

Shelley Circle, 51, 65–66, 77, 152, 171, 243n12

"Sigh for Autumn Rain, A" (Li Yuyang), 107–8

Simond, Louis, 154, 154–55

Sinde (River), 73–76

Singapore, 24

Slave Felony Act, 239n21

slavery, 15–16, 23–24, 79, 95, 111, 129, 211, *223*, 239n21, 240n33

Smith, Adam, 30

Smithsonian Institution, 42–43

snow, 1; agriculture and, 60–61; Alps and, 151–52, 154, 157, 163; America and, 200–203, 205, 210, 215; Arctic and, 130, 144, 147; Howard and, 56–58; ice core analysis and, 37, 43–44,

48, 241n12; Pinatubo and, 42; red, 67; summer, 98; Switzerland and, 45–46; Yunnan and, 98, 102, 105, 115

South Pole, *38*, 48

starvation, 70, 148, 233–34, 239n14; America and, 138, 205, 226; "Bread or Blood" slogan and, 61; Byron on, 69; children and, 5, 9, 23, 25, 63–65; eating boots and, 144; Franklin and, 144–45; glaciers and, 155; Germany and, 61; Ireland and, 172–73, 181, 187, 189, 194, 198; Lord Liverpool's response to, 62; Pentrich Revolution and, 61–62; refugees and, 63–64, 98, 185, 187–89; Sumbawa and, 19, 23–24; Swiss bakers and, 62; Switzerland and, 61–64; Yunnan and, 105, *112*, 113–14

Statham, James, 84–85

Stommel, Elizabeth, 237n5

Stommel, Henry, 237n5

storms, 1; America and, 200–205, 227; Annsville and, 200–201; Arctic Oscillation and, 50–51; Bengal and, 80–83, 99; Byron and, *68*, 77; coldness of 1810s and, 39; Coleridge and, 70; cyclones, 50, *80*; floods and, 56 (*see also* floods); *Frankenstein* and, 9, 45, 52–55, 59, 91, 107, 174, 227–28; hail and, 58–59, 83, 163, 179, 203, 236; heat and moisture transfer and, 50; Howard and, 55–59; hurricanes, 18, 32, 58; Iceland and, 50–51; Ireland and, 176, *177*, 179; lightning and, 53–54, 57–59, 115, 179; London and, 60; Masters and, 54; meteorology and, 54 (*see also* meteorology); monsoons, 22, 78–84, 88–91, 99, 102; northeasters, 46; Plymouth tree and, 59–60; Scotland and, 51; Shelleys and, 9, 45, 52–55, 59, 77, 130; Switzerland and, 46–47, 51–53; Tambora-driven, 50–51, 55–56; thunderstorms, 50–51, 54, 59, 203; Yunnan and, 115

stratosphere, 3, 8, *19*, *38*, 43, 48–50, *52*, 67, 78–79, 82, 90, 115, 230

sugar, 27

Sulawesi (Celebes), 14, *23*

sulfate aerosols, 3, 37, *38*, 48, 78–79, 81, 121, 123, 141, 230–31 230

sulfur, 5, 6, 43–44, 48, *52*, 79

sulfuric acid, *52*, 79

Sumbawa, 5, 37; agriculture and, 13, 29; ash and, 12, 15–25, 30; Bali and, *14*, 19, 22–24; climate change and, 24; deforestation of, 24–25; dry season of, 13; Dutch and, 13–15, 20, 24–25, 27, 29–30, 62; economic devastation of, 23–25; famine and, 24, 30; Flores and, 14, 18, *23*; horse breeders of, 14; Java Trench and, 8;

Koteh and, 13, 17–19; Lombok and, 8, *14*, 22–23, 25; pirates and, 15; poetry and, 17, 37; poverty and, 25; rice and, 13–14, 16–20, 23–24, 31; Sanggar Peninsula and, 7, 13–24, 170; settlement of, 14–15; Sulawesi (Celebes) and, 14, 23; Sumba and, 8; Tambora's death toll on, 17, 19–20, 30; tsunamis and, 18–19, 22; vassal system of, 14–15; *zaman hujan au* (time of the ash rain) and, 24–25

summer of 1816: as "Year of the Beggar," 9, 59, 61; "Eighteen-Hundred-and-Froze-to-Death" and, 9, 128, 200–206, 217; "Franken-stein weather" and, 91, 107, 174, 227–28; June 6 storm and, 200–203; as "Year with-out a Summer," 9–10, 32, 39, 41, 55, 58, 60, 127, 174, 176, *177*, 199–200, 203, 225

Sunda arc, 8

sunsets, 2, 97

sunspots, 40–41, 70

superstition, 16, 25–26, 55, 74–75, 161

Surrey Gardens, 35, 230

Swiss Society of Natural Sciences, 157, 162, 166–67

Switzerland: agriculture and, 62, 64; Alps and, 151 (*see also* Alps); avalanches and, 46, 152; famine and, 62–66, 69; floods in, 58; Shel-leys and, 45–46, 51–55, 57, 62, 65, 151; storms and, 46–47, 51–53; Swiss Society of Natural Sciences, 157, 162, 166–67; Tambora's effect on, 48, 50–51, 57, 61–67; Villa Diodati and, 46, 51, 67, *68*

syphilis, 183

Tambora: Alpine effects of, 45–46, 150–51, 155–56, 159, 162–63, 165, 168–70; American effects of, 71, 199–203, *204*, 216–27; Arctic ef-fects of, 121–23, 132, 135, 138–41, 146, 149; ash and, 48 (*see also* ash); background climate for, 39; caldera of, 5–7, 11, 21; as cautionary tale, 234; climate change and, 233 (*see also* climate change); Crawfurd on, 15; England effects of, 57–60; evacuations and, 5, 6, 8; first rumblings of, 16–17; first victim of, 17; flora/fauna effects of, 229; "Frankenstein weather" resulting from, 91, 107, 174, 227–28; German effects of, 61; hydrologic cycle dis-rupted by, 139–40; immediate deaths from, 17, 19–20; implosion of, 21; Indian effects of, 71–72, 78–83, 90–91, 96; Irish effects of, 71, 171–74, 178–81, 186, 189–90, 193–95, 197–98; Koteh and, 13, 17–19; lost kingdom of, 25; monsoon effects of, 78–84, 88–91; panic of 1819 and, 216–26; pumice hail of, 17, 21–22;

pyroclastic flows and, 18–20, 22, 170; Sang-gar Peninsula and, 7, 13–24, 170; scale of eruption of, 42–44; sound of, 21–22; storms and, 50–51, 55–56 (*see also* storms); strato-spheric effects and, 48, 78, 82, 90; sulfate aerosols and, 3, 37, *38*, 48, 78–79, 81, 121, 123, 141, 230–31 230; Sumbawan deaths from, 17, 19–20, 30; Sunda arc and, 8; superstition and, 16, 55; Swiss effects of, 48, 50–51, 57, 61–67; Teluk Saleh and, 7; wind and, 17–19, 22, 48, 55, 61, 78–79, 83, 87, 107–8, 111; Yunnan effects of, 98, 102–8, 111–13, 115, 117–20

Teluk Saleh, 7

temperature, 9; atmospheric pressure and, 49; average, 39, 178, 205, 209, 217; Buffon on, 209, 214; cholera and, 90; cycles of, 166; febrile body, 184; glaciers and, 152, 157, 159–60; gra-dient of, 50, 80, 81, 203; higher, 18, 88, 89, 138, 140; Holocene Period and, 42, 44, 232–33, 241n12; Howard's observations and, 58; jet streams and, 50, 203; Land Office records and, 227; latitude and, 137, 209; lower, *38*, 39, 41, 50, 55–56, 58, 78, 79, 89, 98, 103–5, 115, 127, 129, 131, 141, 150–51, 178, 200–203, 205, 216; measuring seawater, 135–37; meridional currents and, 48–50, 82, 137–38, 249n25; moderate, 42, 49, 100, 102, 137, 212, 215, 226, 228, 230; ocean, 135–38; raw, 49, 79; regional, 207; rice crops and, 105; sulfate aerosols and, 230; wind and, 49 (*see also* wind)

Tenerife, 34

Tennyson, 146

Thailand, 91, 119

Theatre de la République, 35–36

thunderstorms, 50–51, 54, 59, 203

Times newspaper, 194

"To Autumn" (Keats), 39

trade winds, 19, 23, 78, 82, 99

transatlantic trade, 64, 138, 210, 217, 222, 224, 226, 228, 237n5

Treaty of Paris, 1

tree-ring studies, 24, 43, 82–83, 217, 248n19

Trevelyan, Charles, 174

tropics, 2, 8, 127, 151, 232; air pressure differen-tials and, 49; Atlantic Meridional Over-turning Circulation (AMOC) and, 137–38, 140, 249n25; disease and, 94; eruption of 1258 and, 41, 43–44, 216; eruption of 1809 and, 37–39, 41–43, 58, 216, 241n12; forests of, 17; heat transfer and, 137–38, 140, 249n25; Little Ice Age and, 37, 39; monsoons and, 22, 78–84, 88–91, 99, 102; rain in, 5, 15; raw

tropics (*cont.*)
surface temperature and, 79; rice and, 103; summer snows and, 98; Volcanic Explosivity Index (VEI) and, 43; volcanoes located in, 2, 41, 43–44 (*see also* volcanoes); wind-based heat transport and, 49
tropopause, 50
Trotter, John, 185
tsunamis: Alpine ice floods compared to, 150–70; Sumbawa and, 18–19, *22*
tuberculosis, 77, 221
Turner, J. M. W. (William), 2–3, *35*, 58, *156*, 174
typhus, 69; Burundi and, 183; Carleton's writings and, 184, 188, 190, 192; clergy and, 190; doctors and, 181, 186, 189–92; etiology for, 183; famine and, 9, 70; Gamble and, 186–87; Harty and, 186, 190–91; Ireland and, 66, 176, 179, 181–93, 198; lice and, 176, 179, 181–87; Napoleon's army and, 183; newspaper media and, 190; pathogen source of, 183; rural population and, 188–89; treatises on, 187, 253n43

unknown volcanic eruptions: of 1258, 41, 43–44, 216; of 1809, 37–39, 41–43, 58, 216, 241n12

Val de Bagnes: Agassiz and, 161, 167–68; Charpentier and, 161–62, 167–68, 251n24; Perraudin and, 160–62, 164, 166–68; Venetz and, 159–68, 170
vampires, 52, *52*–53, 66
Vanautu, 37
Venetz, Ignace: climate change and, 166–67; engineering talents of, 157, *158*, 161–62, 164, 167; Giétro and, *158*, 159–61, 163–65, 170; glacier studies of, 157–58, 160–68, 170; Swiss Society paper of, 166–67; Val de Bagnes and, 159–68, 170
Vesuvius, Mount, 18, 33–36, 44, 153
Vietnam, 14, 98, 117
Villa Diodati, 46, 51, 67, *68*
Volcanic Explosivity Index (VEI), 43
volcanoes: ash and, 2 (*see also* ash); calderas and, 5–7, 11, 21; climate change and, 5 (*see also* climate change); climate impacts of recent, 231; coldness of 1810s and, 36–42; ejecta of, 2, 48; eruption of 1258 and, 41, 43–44, 216; eruption of 1809 and, 37–39, 41–43, 58, 216, 241n12; evacuations and, 5, 6, 8, 163; fluorine and, 24, 48; Fuji, 8;

Huaynaputina, 44; Ice–Core Volcano Index (IVI) and, 43–44; Iceland and, 1–2, 7, 34, 40, 43; Katmai, 42; Krakatau, 8, 37, 41–42, 44, *53*; Kuwae, 37, 44; Laki, 2, 43; magma and, 6, 17–21, 43; Pinatubo, 42, 48, *52*, 138–39, 241n22; impacts on precipitation, 49, 217; pumice and, 17, 21–22, 33; pyroclastic flows and, 18–20, 22, 170; Ring of Fire and, 8; St. Helens, 8, 43; Santorini, 44; stratospheric effects of, 3, 8, *19*, 43, 48, 50, *52*, 67, 79, 90, 115, 230; sulfate aerosols and, 3, 37, *38*, 48, 78–79, 81, 121, 123, 141, 230; sulfur and, 5, 6, 43, 48, *52*, 79; symbolism of, 35–36; Tambora and, 2 (*see also* Tambora); Tenerife, 34; unknown eruptions of, 37–39, 41–44, 58, 216, 241n12; Vesuvius, 18, 33–36, 44, 153; visual entertainment and, 34–35; Volcanic Explosivity Index (VEI) and, 43
volcanologists, 6, 10, 26, 33–34, 43, 235, 238n10
Volney, Constantin-François, 215
von Clausewitz, Carl, 61
von Drais, Karl, 229
von Kotzebue, Otto, 128, 131, 141

Wales, 62, 116
Wang Can, 11
War of Independence, 1
Waterloo, 70, 116, 148, 180
Weymouth Bay (Constable), *50*
whaling, 125, 127, 135
wheat: Alps and, 164; America and, 203, 205–6, 214, 221, *222*, 226; Ireland and, 176; Yunnan and, 99, 102, 105, 110, 115, 119
Williams, Samuel, 215, *215*–16
wind: Alps and, 45–46; Arctic and, 140; America and, 203, 215–16; Bengal and, 78–79, 83, 87, 89; climate and, 49–51, 150, 173, 178; dark, 104; hurricane-strength, 17–18, 58; Ireland and, 173, 178; Switzerland and, 53; Tambora and, 17–19, *22*, 48, 55, 61, 78–79, 83, 87, 107–8, 111; trade, 19, *23*, 78, 82, 99; Yunnan and, 99–100, 102, 104, 107–8, 111
World War I, 149

Yang Yuda, 247n3
"Year of the Beggar," 9, 59, 61
"Year without a Summer," 9–10, 60; America and, 199–200, 203, 225; Banks and, 127; Howard and, 58 Ireland and, 174, 176, *177*; legends of, 55; Little Ice Age and, 41; Mackenzie and, 39; Sumbawa and, 32

Yeats, W. B., 11
yellow fever, 183
Yunnan, 5; agriculture and, 99–103, 107–8,
    117–20; climate change and, 119–20; climate
    of, 98–101, 102–5; Confucianism and, 98,
    106–7, 109–10, 114, 120; famine and, 102–20;
    Han Chinese expansion and, 100–201, 103;
    Himalayan plateau and, 99, 102, 115; Jiaqing
    and, 109–10; as "Land of Eternal Spring,"
    100–1; Li Yuyang and, 11, 106–15, 120, 236;
    meaning of name, 100; Mekong delta and,
    98, 102, 119; meteorology and, 97–99, 104–5,
    109; monsoons and, 99, 102; opium and,
    115–20; poetry and, 11, 106–15, 120; poverty
    and, 107; Qing dynasty and, 100, 101, 105–13,
    117, 120; rice and, 71, 99, 101–5, 107–11, 113, 119;
    storms and, 115; Tambora's effects on, 98,
    102–8, 111–13, 115, 117–20; wheat and, 99, 102,
    105, 110, 115, 119; winds of, 99–100, 102, 104,
    107–8, 111

zaman hujan au (time of the ash rain), 24–25